THE FRAGILE WISDOM

THE FRAGILE WISDOM

AN EVOLUTIONARY VIEW ON WOMEN'S BIOLOGY AND HEALTH

Grazyna Jasienska

HARVARD UNIVERSITY PRESS

Cambridge, Massachusetts

London, England

2013

Library of Congress Cataloging-in-Publication Data

Jasienska, Grazyna.

The fragile wisdom : an evolutionary view on women's biology and health /
Grazyna Jasienska.

pages   cm

Includes bibliographical references and index.

ISBN 978-0-674-04712-9 (alk. paper)

1. Women—Health and hygiene.   2. Women—Physiology.
3. Reproductive health.   4. Women—Evolution.   I. Title.

RA778.J43   2013

613'.04244—dc23   2012025259

To Michał and Adam

for making my life trade-offs free

CONTENTS

THE FRAGILE WISDOM

Why It Is So Difficult to Be Perfectly Healthy

Many women put a lot of effort into following what they are told is a healthy lifestyle. They do not smoke, eliminate fatty foods from their diets, cut back on sweets, take stairs instead of elevators, and walk to work. When they get sick with breast cancer, heart disease, or osteoporosis, they often blame themselves and ask, "What did I do wrong?" We hear so much about how different diseases can be prevented by a "healthy" way of living that when we become ill we often believe that it happened because we somehow failed in our attempts to follow all the recommendations. I hope that this book will show that people should not blame themselves for becoming ill. Of course, there are behaviors that without a doubt are bad for us. Smoking increases the risk of many diseases, and not smoking or giving it up results, for an average person, in much better health and an extended life. What other lifestyle decisions would have a similar effect? Most people would say exercise and eating right. That is true, but in fact we do not know for sure what exercise and eating right really mean when it comes to providing concrete advice to a specific individual. One of the themes of this book is an attempt to understand why specific recommendations about physical activity and diet are so difficult to formulate.

Some researchers think that the key to understanding what is good for us when it comes to a recommended lifestyle is greater knowledge about the physical activity and diet of our ancestors during evolutionary times (i.e., more than 10,000 years ago). This is clearly a good starting point, but the question remains as to whether what was good for our ancestors is still appropriate for us. We cannot replicate their past life, but if we can pick just some aspects of an evolutionary lifestyle and incorporate them into our lives, would that be enough to prevent diseases?

For women, modern life is associated with major changes in reproductive patterns in comparison with what our evolutionary female ancestors experienced. Some of the components of this modern reproductive pattern can be modified: women can change their lifestyle relatively easily. For example, they can increase their level of physical activity and, as a result, reduce the levels of their own reproductive hormones and thus achieve a lower chance of breast cancer. But for modern women, modifying other components of their reproductive lives, such as early sexual maturation, low fertility, or a shorter period of breastfeeding, is very difficult or impossible. This book does not provide an easy to follow prescription for disease prevention. In fact, this book does not even attempt to offer a prescription at all. This is a book for people who would like to understand why following recommendations for a healthy lifestyle does not always work. This book is for people who are interested in tracing and exposing numerous evolutionary prerequisites and underlying themes crucial for human biology and health. This is also a book for people who do not mind crossing boundaries of different scientific disciplines, and are not afraid of some speculative thinking (or, in scientific terminology, of building and considering hypotheses).

Is the Human Body Poorly Designed?

"We must be really poorly designed if such physical exertion is necessary to stay healthy," complained an article about jogging in a women's magazine. They got it all wrong: humans are not badly designed. In fact, we are not more poorly put together than most other living organisms. The problem lies elsewhere: our design was never meant to ensure a low risk of heart disease, diabetes, or breast cancer. Health and evolutionary fitness are not synonymous. Our physiology, anatomy, even behavior have evolved to facilitate reproduction—passing our genes to the next generation. Being healthy is an important part of that process, as individuals must survive until the age of reproductive maturation. In a long-lived species such as ours, we should live well past that age so that more than one child can be born and raised. However, natural selection, the most important of the processes responsible for shaping the functional features of human anatomy and physiology, does not care much about health and survival in the postreproductive years. In fact, any trait, no matter how detrimental to health and survival in the postreproductive period, is promoted by natural selection if it increases the chance of having offspring at a younger

2

age. One may think about this as a conflict of interest between reproduction at an early age and health at an older age. You cannot have your cake and eat it, too.

Ways of looking at the functioning of the human organism have changed since 1932 when Walter B. Cannon, a Harvard physiologist, published his classic book *The Wisdom of the Body,* but the notion that the human body "knows" how to deal with external challenges is still powerful. Many people believe that humans have an innate ability to make the right "healthy" lifestyle choices. Several recent popular books have used the phrase "the wisdom of the body" on their covers, and many others have used exactly this kind of approach when talking about dietary choices, exercise, or other tactics that their readers supposedly should implement to preserve or improve their health.

In this book I am not going to vehemently argue against the notion of the wisdom of the body, but I will try to show that in many respects this wisdom is imperfect and rather fragile. Some problems faced not only by humans but by most other species as well are created by the existence of inevitable trade-offs, confounded by the unavoidable costs of having offspring, and they are frequently magnified or imposed by adverse developmental conditions. Other problems more unique to humans result from a mismatch between our evolved and present lifestyles, and they are often further complicated by the demands imposed by culture.

A note about language is warranted here. In this book I will use phrases such as "the mother keeps in mind," "the fetus predicts," "the fetus assumes," or "the body decides," which are typical shortcut expressions used by many people in the evolutionary field. We know that the fetus does not really "know," not in the sense of being aware of making certain physiological decisions. Rather, it "behaves or responds physiologically as if it knew." If we unpack this statement further, we admit that an organism uses responses that are built into a coherent repertoire by the process of natural selection.

Allocations and Trade-Offs

Everything in life has a cost. An individual allocates energy from digested food to physiological maintenance and physical activity, and stores what is left as fat. When this individual sexually matures, she needs to make an allocation to one more process: reproduction. If resources are limited, energy used to support reproductive processes will result in less energy allocated to other processes. This means other processes will not receive as much energy as they optimally require, and in consequence the organism will not function as well as it would if it were "celibate" or sterile.

Life-history theory sees the individual organism as a set of particular traits and events occurring during its life. Size at birth, rate of growth, age of sexual maturation, and number and size of offspring are examples of traits and events important for individual life history. How these traits develop and when these particular events occur depend on the availability of metabolic energy and on decisions of how to allocate energy among competing traits or events. What is most important here is that energy used for one purpose cannot be used for another purpose. If energy is limited, as it usually is, the organism has to make decisions about which processes are the most important during a given stage of life. These processes will be supported, and others will not get enough energy. A pregnant adolescent girl will not grow as rapidly as her age counterpart who is not yet reproducing, and her final adult size will be smaller. She will also give birth to a smaller baby than an older woman would: she allocates a lot of energy to the developing fetus, while at the same time she must still allocate some to support her own growth.

During her lifetime, a woman has a finite quantity of resources she can allocate to support reproduction, which includes pregnancy, lactation, and parental care. Her lifetime reproductive success is best served by careful allocation of energy to each reproductive event. This means that a single offspring will not receive as much energy or nutrients as it would like to have, especially when maternal nutritional condition is poor, because the mother must take into account her future prospects for continuous reproduction and the need to pay continuous expenses. Investing too much in a single child may lead to the deterioration of the mother's condition and may negatively affect her future reproduction.

An example of the eternal dilemma faced by living organisms is the trade-off between reproduction and fighting infection. A nonreproducing individual may have sufficient resources to run its immune function at a level to keep infections in check. That immune system may function at a suboptimal level in a woman who is currently nursing a child or a man who is actively competing with other men for access to mates. Likewise, an individual living in an environment with a high parasitic load may not reproduce as well as an individual from a pathogen-free environment. Such short-term trade-offs may have long-term consequences; they may result in conflict between reproduction and health in later years, or even between reproduction and life span. Consequently, individuals paying higher reproductive costs may live shorter lives.

Living organisms develop during a complex interaction between genetic and environmental influences. As if this were not complicated enough, genes themselves work in such a way that usually one gene affects more than one trait. This phenomenon is called *pleiotropy*. When the same gene has a positive effect on one trait but a detrimental effect on another, we call this *antagonistic pleiotropy*. For example, the effect of estrogens (i.e., steroid hormones such as estradiol, one of the hormones produced during the menstrual cycle) in women can be seen as antagonistic. High levels of estrogens aid reproduction because they are directly related to conception, but exposure to high levels of these hormones also increases mortality from estrogen-dependent cancers such as breast cancer. To complicate matters further, estrogens are involved in additional trade-offs in the postreproductive years: high estrogen levels may increase the risk of cancers but may lower the risk of other late-age diseases such as osteoporosis, heart disease, and possibly depression and dementia.

Such life-history trade-offs explain, to a certain extent, why it is difficult to stay healthy, and why there is little anyone can do to change that. These trade-offs are a fact of life. As Steven Stearns, Randolph Nesse, and David Haig wrote recently, "One of the most useful generalizations evolution offers to medicine is a vision of the body as a bundle of trade-offs. No trait is perfect. Every trait could be better, but making it better would make something else worse" (2008, 11).

Five decades ago, evolutionary ecologists began to focus on the dilemmas of priorities, allocations, and trade-offs, and life-history theory is now a blooming branch of evolutionary biology. I will not refer to the theoretical developments in this area; instead, I will use the approach of a humble empiricist who is trying to capture the complexity of the problem in commonsense terms. Humility does not imply simplemindedness but rather readiness to accept that the era of pop medicine and pop public health is over. Simple take-home messages and one-line recommendations do not exist.

Poor Fit between Stone Age Physiology and the Modern Lifestyle

Human physiology, anatomy, and behavior evolved over a long period of time. Many features were inherited from our primate ancestors, and many others were shaped by the process of natural selection after the human species had begun its separate evolutionary journey several million years ago. The success of a newly evolved trait depended on its fit with the environment, and as the environment changed, some previously advantageous traits lost their ability to increase the reproductive success or biological fitness of individuals.

Some diseases that occur in later life also result from a poor fit between the environment for which humans were (evolutionarily) designed and the environment in which modern people live. The main problem here is that the lifestyle most people lead differs dramatically from the lifestyle of our evolutionary ancestors. Our ancestors lived for more than 90 percent of their evolutionary history as hunter-gatherers, a lifestyle that involved a very different diet, very different patterns of physical activity, and a very different web of social interactions. For women, it also meant a very different reproductive life: late maturation, fewer menstrual cycles, several widely spaced pregnancies, and long-lasting breastfeeding.

Most of human evolutionary adaptations are believed to have been shaped during that time, which means modern humans are adapted to the lifestyle of hunters and gatherers. This subsistence strategy changed relatively recently, no more than 14,000 years ago, when people discovered that they could successfully grow their own food instead of relying on nature's supplies. Agriculture slowly spread all over the world and brought substantial changes to the human lifestyle. Not only did the diet become very different, but there were also changes in physical activity, reproduction, social interactions, and the types and prevalence of diseases. Urbanization and more recently industrial society brought further changes, making our modern lifestyle even more divergent from the lifestyle of our hunter-gatherer ancestors.

As the lifestyle changed, human physiology was left behind because 14,000 years have not been enough for major evolutionary changes in physiology to be built into the human repertoire. Of course, some changes have occurred since the origin of agriculture. For example, in some pastoralist societies, the ability to digest milk evolved for adults, an ability that is normally lost in early childhood. It is estimated that it took 8,000 years—about 325 generations—for the genetic mutation that allowed adults to digest milk to increase in frequency in the population from 1 percent to 90 percent (Stearns, Nesse, and Haig 2008). Eight thousand years is a very short time for a feature or a trait to increase its frequency so substantially, which occurs only for traits that bring a significant advantage

6

to the individuals expressing them and only for traits that are coded by a single, dominant mutation (most traits have much more complex genetic backgrounds). Several other similarly small changes clearly happened in response to the environmental pressures of agricultural and modern life, but it is generally believed that most of the physiological and metabolic processes that keep modern humans going are exactly the same as they were in our hunter-gatherer ancestors of the Paleolithic era.

Poor Fit between Fetal and Adult Environments

Just as a lack of fit between past evolutionary environments and modern, Western lifestyles can be blamed for many human diseases, a mismatch on a much shorter time scale is also responsible for an increased risk of diseases. A fetus developing in a poor maternal environment will experience a higher risk of several diseases many years later during adult life, especially when the adult environment is rich in resources. Hypotheses claim that this lack of fit between very early life and adult conditions constitutes the worst possible scenario for the life prospects of the individual. The physiology and metabolism of such an individual are predisposed for problems with insulin metabolism and high blood pressure, which consequently often leads to diabetes or heart disease. If the adult environment better matches the fetal environment in its poor or rich quality, these health problems are less likely to develop.

We do not fully understand why an individual developing in a poor maternal environment has a permanently affected physiology and metabolism. An obvious explanation would be that in such a poor environment the fetus does not receive sufficient energy and nutrients necessary for optimal development; as a result of these developmental constraints, its physiology is damaged and does not develop correctly. Another explanation is that the period of fetal development constitutes preparation for the future. The fetus develops in the way that will be optimal in a future environment. How does the fetus know in what kind of future environment it is going to live? It does not know for sure, but it can make reasonable predictions based on the conditions experienced during early development. There are no other "data" available for the fetus to analyze. If the fetus develops in poor conditions, its predictions for the future are not very optimistic—it "assumes" that future life will be equally bad. Consequently, its physiology and metabolism develop, as some hypotheses suggest, in a way that would do a relatively good job in those expected poor conditions. So, for example, its body size may be smaller because a small body needs less energy, and its ability to store energy may be better, as energy stores will be especially important in an environment of poor quality.

But sometimes the fetus makes wrong predictions. If the adult environment turns out to be abundant in energy, all those metabolic and physiological adjustments made in utero in preparation for an energy-constricted life are not only unnecessary but are actually harmful. The harm manifests for adults, mostly those in their postreproductive years, when diseases such as heart disease or adult-onset diabetes usually occur. We do not know whether this changed physiology is harmful or beneficial for younger, reproductive-age individuals when facing a mismatch between their fetal environment and the environment of their early adulthood.

Did an environmental mismatch occur often during human evolution, or are we witnessing the emergence of a new phenomenon? It is probably the latter. One may safely assume that large differences in environmental conditions from fetal to adult years, to the extent experienced in modern times, were unlikely during human evolutionary history. You were born poor, you grew up poor, and you died poor. In particular, the adult environment was never as abundant in energy as it is today. Consequently, the existence of such an environmental mismatch was not a cause of consistent, significant selective pressure. In other words, it was never a challenge to be solved by evolving the necessary adaptations by means of natural selection.

Nevertheless, maternal condition during the development of the fetus is of crucial importance for the future condition of the individual. Fetuses deprived of energy and nutrients are born small, which is a predictor of mortality and morbidity, especially during infancy. Such individuals also have an increased risk of diseases much later in life. If the consequences of poor nutrition during fetal development are so dire, why would the mother not invest more in the developing offspring? Why did humans not evolve mechanisms that would allow the mother to pass more energy to the developing child?

Such mechanisms, in fact, do exist, and mothers do make some physiological sacrifices during an energy-deprived pregnancy or lactation. But the mother will not produce a big, healthy child if it costs her too much. This is another life-history trade-off, namely, between current and future reproduction. Such trade-offs are responsible, to some extent, for the low birth weight of babies whose mothers are nutritionally deprived.

Biology and Culture

During the course of human evolution, environmental conditions affected human life history just as they did any other species. But in humans, unlike in most other species, culture had and still has an important influence on life history. Two ancient, major cultural innovations, making stone tools and using fire, immensely extended the dietary breadth of our species in comparison with ancestral diets. Today, cultural practices and beliefs are very rich and diverse, and how they affect human life history is a complicated issue, well beyond the scope of this book. But the role of culture is important, because cultural practices and customs may interfere with biological trade-offs. Obvious examples to consider are associated with medical care and technology: babies with very low birth weight, who clearly did not use many maternal resources during pregnancy, can now survive, and infant formula frees the mother's physiology from the costs of lactation.

In populations with more traditional lifestyles, postpartum sex taboos preventing both partners (but mostly the woman) from having sexual intercourse for some time after delivering the baby may be important for keeping interbirth intervals sufficiently long. That role during human evolution was successfully served by breastfeeding among hunting-gathering mothers. Prolonged, high-frequency nursing together with the relatively poor nutritional status of the mother extended postpartum infertility.

Once humans adopted the agricultural lifestyle, for several reasons breastfeeding became less efficient for preventing new pregnancies, and postpartum taboos became important cultural strategies to preserve the biological well-being of mothers and babies. We may think of them as purely cultural phenomena, adjusted through accumulated knowledge passed across generations. Such strategies serve adaptive functions, but they should not be confused with biological, naturally selected adaptations.

It is important to stress at the beginning of this book that, of course, biology is not destiny. People can make conscious choices that can go against biological laws. According to life-history theory, mothers should optimize lifetime reproduction and not invest everything into a single

9

child. Many examples described in this book suggest that this is indeed how female physiology works. Women's bodies sacrifice the condition of the current child in utero if that action will increase their chances for future reproduction. But, while female physiology has evolved to work that way, women can make individual decisions that do not follow this evolved strategy. For example, during the sixth month of her pregnancy, Anna, a twenty-four-year-old woman in Poland, received a diagnosis of skin cancer, which had already spread to her lungs (Krzyk and Steinhagen 2007). Doctors told her that if chemotherapy was started immediately she would have a very good chance of overcoming the disease, but the treatment could harm her child. For the sake of the child, she decided to forgo treatment. She gave birth to a healthy baby boy, Oskar, and then began therapy, but the pregnancy had accelerated the progress of her cancer, and surgery was no longer possible. All other treatments were unsuccessful, and Anna died when Oskar was just six months old.

Conscious choices are often not enough to outwit biology. Eating a lot during pregnancy will have a limited impact on the child's birth weight. This would be the case no matter how conscientiously a mother supplements her diet with additional calories while pregnant, no matter how much she wants her child to be born big and healthy. However, additional dietary intake in pregnancy does not go unnoticed by the maternal physiology, as it could potentially shorten the amount of time she needs to conceive again.

What to Expect from This Book

This is not a self-help book, and it will not provide a simple solution to how to lead a long and healthy life. I will not attempt to provide a single take-home message to help people avoid diseases or design a public health program with the power to reduce the disease burden at the population level. Further, instead of spreading the main concepts too thin, I have decided to focus on selected examples.

This is a book about health. I will be mostly talking about the reproductive biology and health of women. What is important here is an evolutionary perspective. It means that I will attempt to convey the challenges faced by women's physiology and health by use of the framework of evolutionary biology and especially life-history theory. This book will be based on concepts, ideas, and findings from the areas of human reproductive ecology, Darwinian medicine, and evolutionary public health.

I will be touching on topics that range from how exercise reduces levels of reproductive hormones, to the prevention of breast cancer, from the history of programs designed to improve child health in France, to their possibly long-lasting impact on the health of the contemporary French population.

Other long-lasting effects of historical life conditions on health will be explored by the hypothesis that the long history of slavery can be held partially responsible for low birth weight among modern African Americans. Interaction between cultural practices and reproductive biology will be exemplified by a discussion of the impact of the dietary changes accompanying agriculture on the genes responsible for coding reproductive hormones and by asking if the procedure of fattening Moor girls in the Sahara has any adaptive biological significance. Talking about women and reproduction requires a look at the costs of having children, and I will discuss potential links between fertility and health as well as between fertility and longevity.

I will examine whether health can be preserved through appropriate diets and physical activity, mostly by asking what the appropriate diets and exercises really are and how we can know. People with low birth weight have an increased risk of cardiovascular disease when they are older, and for them any preventive measures would be especially important. For women, a low birth weight leads to a higher risk of cardiovascular problems, but a high birth weight is far more dangerous in terms of breast cancer risk. The amount and intensity of exercise that is necessary to keep the risk of many diseases low may vary, depending on the person's birth weight or weight gain during childhood.

One could argue that health prevention has already had its day in court. Programs promoting health have operated for several decades and clearly have had limited success, as the rates of many lifestyle diseases have not declined. But that was old-fashioned prevention—modern prevention must be based on knowledge about human evolutionary history and on an understanding of the principles of life-history theory. Modern prevention must look at the entire life course of the individual, where all stages of life are connected and affect each other. Thus, knowledge about early conditions from the fetal or childhood period can help predict an individual's future health. Modern prevention must be individualized—custom fitted to an individual's early experiences and reproductive history.

Molecular genetic technologies have created high hopes for fixing the health problems of modern populations. But my view on this is that of a skeptic: I think prevention, not gene therapy, is still the key for reducing the risk of modern civilization's major diseases. We cannot sit back and rely on genetic laboratories. We must take responsibility for our own health.

I argue in this book that without evolutionary thinking our understanding of our own health is superficial, and consequently our health programs are likely to backfire. Theodosius Dobzhansky was right that nothing in biology makes sense except in the light of evolution, but for evolutionary thinking to be useful we must begin to appreciate how entangled, complex, and multifaceted a phenomenon our health truly is.

Many aspects of women's biology and health depend on reproductive hormones: not only fertility, but also psychological well-being and diseases that are major health risks for women today, such as breast cancer and osteoporosis. Levels of the reproductive hormones, estrogens and progesterone, depend on genes and the conditions we experienced during fetal and childhood development. Levels of these hormones are not stable throughout life; they change not only with age but with lifestyle. Chapter 1 discusses the ideas and research findings of human reproductive ecology, the area of science that studies variation among people in reproductive function and attempts to identify its causes.

1

IF REPRODUCTIVE HORMONES ARE

SO IMPORTANT, WHY IS THERE

SO MUCH VARIATION?

When it comes to the area of fertility, many people, including some health professionals, believe that women of reproductive age can be divided into two groups: those who are fertile, and those who are not. The group with fertility problems is very diverse, as medical science today can identify numerous anatomical, genetic, physiological, and metabolic disorders leading to fertility impairments. The group of healthy, fertile women is considered much less interesting, especially if they have regular menstrual cycles, because they can conceive children without assistance, which places them beyond the interest of medical professionals specializing in reproductive health. What the medical field is not taking into account, however, is that healthy, fertile women who have no reproductive disorders still differ in many ways in their reproductive physiology. This variation is interesting and important for several reasons.

First, the variation contributes to differences in fertility (i.e., the number of children women have). It is not true that all healthy women can have a similarly large number of children so long as they do nothing to control their fertility. Women from populations with natural fertility—that is, fertility achieved without contraception or other methods of family planning—show significant variation in the number of children they have during their reproductive lives (Campbell and Wood 1988; Bentley, Goldberg, and Jasienska 1993; Bentley, Jasienska, and Goldberg 1993). The underlying causes of such variations are, to a large extent, biological: they result from differences in ovarian function. These differences, in turn, result mainly from the impact of the environment in which women live, especially from the factors that determine how much metabolic energy women have and can allocate to reproduction.

Variation in reproductive function among healthy women in response to the environment—which in humans we can call "lifestyle"—is believed to reflect the ability of reproductive physiology to react in a way that is biologically appropriate, or, in the language of evolutionary biologists, *adaptive*. Peter Ellison (1990, 2003b) has stressed that temporarily suppressed reproductive function, and thus a temporarily reduced ability to conceive, is an adaptive response to preserve the long-term ability of a mother to bear children in the future. This aspect of variation in ovarian function is particularly interesting for those who study evolutionary biology and life-history theory, for short-term suppression may affect reproductive success, which is how biological fitness is measured.

Appreciating the existence of variation in reproductive function is useful for many other theoretical areas as well: for example, it may help explain certain demographic trends. It is especially important for all aspects of health and disease prevention in women. The impact of variation in reproductive function is obvious in terms of fertility, but circulating levels of reproductive steroid hormones are also important determinants of other aspects of health and disease, which gives knowledge about variation in ovarian function practical implications for disease prevention. Extensive research notwithstanding, variation in the incidence of breast cancer and other hormone-dependent cancers (such as uterine and ovarian cancers) is only to some extent explained by the established risk factors for these diseases. Reproductive steroid hormones are important players in the development and progress of these cancers, and knowledge of a woman's lifetime exposure to these hormones may help to predict her risk (Figure 1.1).

Levels of hormones are fortunately not set for life. They can be manipulated not only by medical intervention but also by the woman herself, often through relatively simple changes in her lifestyle. Therefore, the risk of hormone-dependent diseases can, theoretically, be manipulated as well. But to do that efficiently, the determinants of hormone levels need to be established and their interactions well understood.

Variations in levels of endogenous hormones—those produced in our body—may be responsible for variations in tolerance of exogenous hormones—those used in contraception, hormone replacement therapies, and medication. Interpopulation variation in response to synthetic, hormonal contraception has been well established (Vitzthum and Ringheim 2005). Gillian Bentley (1994) has suggested that tolerance of high doses of contraceptive steroids may depend on a women's own endogenous levels, so the recommended standard doses are much too high for women from many non-Western populations. Such high doses are not

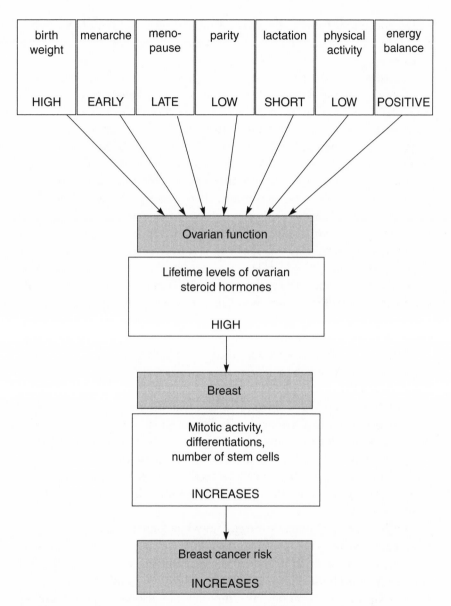

Figure 1.1. Most factors that increase the risk of breast cancer contribute to the elevated lifetime levels of ovarian steroid hormones.

only unnecessary for the effectiveness of contraception but often are not well tolerated and may be associated with unnecessarily elevated health risks.

The severity of perimenopausal symptoms (hot flashes, sleep problems, anxiety, irritability, and depression occurring before and during the menopausal transition when levels of reproductive hormones substantially decline) varies among women from different countries. Such differences may also be explained by variation in hormone levels produced throughout life in menstrual cycles. Women whose physiology is "used to" high levels of ovarian steroids may experience more severe problems when ovarian production diminishes than women who were never exposed to such high levels during their reproductive years.

Women from Japan and mainland China report some menopausal symptoms—in particular, hot flashes—less frequently than women from Canada and the United States (Kaufert et al. 1986; McKinlay and McKinlay 1986; Lock 1993; Ikeda et al. 2005; Shea 2006). Perhaps cultural customs prohibit Asian women from complaining, and perhaps a diet rich in soy products alleviates these problems, but the role of differences in lifetime hormonal exposure is clearly worth further investigation. In a multiethnic study conducted in the United States, Chinese and Japanese participants had lower serum estradiol concentrations than Caucasian, Hispanic, and African American women (Lasley et al. 2002; Randolph et al. 2003). Women in rural China have lower average estradiol levels than British women in the United Kingdom (Key et al. 1990), and Chinese women from Shanghai have about 20 percent lower estradiol concentrations than white women from Los Angeles (Bernstein et al. 1990). This variation in hormone levels seems to correspond to differences in the severity of some menopausal symptoms observed among these populations.

It is important to note, however, that while there is a great diversity in menopausal symptoms in women from different populations, it is not at all clear that women from industrialized countries experience a higher frequency of such symptoms (Obermeyer 2000). Considerable variation in the frequency of menopausal symptoms can also be observed among different Asian populations. The Pan-Asia Menopause Study, which compared symptoms associated with menopause among women from nine populations, including China, Hong Kong, Indonesia, Korea, Malaysia, Pakistan, Philippines, Singapore, Taiwan, Thailand, and Vietnam, found that hot flashes were reported by only 5 percent of Indonesian women,

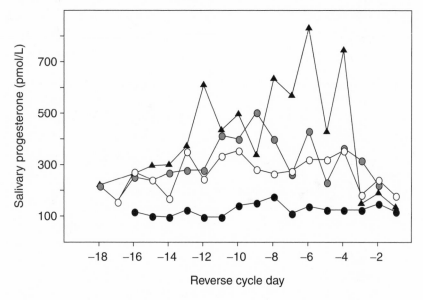

Figure 1.2. Interindividual variation in progesterone profiles: four healthy, regularly menstruating women of comparable age from the same population.

versus 47 percent of Korean women and 100 percent of Vietnamese women (Haines et al. 2005). These results need further verification because the Pan-Asia Menopause Study had only small sample sizes in some of the investigated populations (which could greatly affect the precision of the estimates of relative frequencies). Further reliable hormonal data for a larger number of populations are needed to test the hypothesis that population variation in menopausal symptoms can, at least partially, be explained by variation in the levels of reproductive steroid hormones in women's menstrual cycles.

Steroid hormones, especially those produced by the ovaries, are important for many aspects of female health, well-being, and reproduction. High levels of ovarian hormones in menstrual cycles are crucial for successful pregnancy and, as such, are important determinants of female reproductive success. However, not all women have similar levels or high levels of these hormones. There are substantial differences in the mean levels of estradiol and progesterone among populations, among women within a single population, and even among menstrual cycles of a single woman (Jasienska and Jasienski 2008). For example, urban women in the United States have progesterone levels that are on the average 65

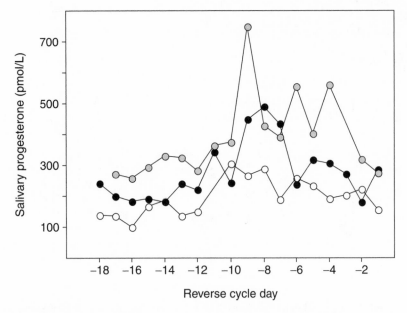

Figure 1.3. Intraindividual variation in progesterone profiles: three consecutive cycles of the same woman.

percent higher than those of women in the Democratic Republic of Congo (Ellison et al. 1993). In a small village in Poland, we found substantial differences in the progesterone levels among women (Figure 1.2), and also among the cycles of a single woman (Figure 1.3). Such high variation is probably caused, in this population, by the seasonality of agricultural workload; progesterone levels are suppressed during the months with intense energy expenditure in comparison with the months with less intense work. Even in urban women from the United States and Great Britain, where lifestyle is much less influenced by seasons than in agricultural populations, progesterone levels can vary substantially from cycle to cycle (Lenton et al. 1983; Sukalich, Lipson, and Ellison 1994; Gann et al. 2001).

Production of Reproductive Hormones: The Menstrual Cycle

The major proportion of reproductive hormones that women produce during their lives comes from processes that occur during menstrual cycles. The body also has other ways to produce these hormones: they can be made by the placenta during pregnancy or converted from other hormones by fat tissue. For most women, especially in Western countries, menstrual cycles are the most important in terms of the amount of hormones produced during a lifetime.

I will briefly describe the events that occur during an ovulatory menstrual cycle. It is important to remember that "real life" cycles may differ from this example: not all cycles are ovulatory, and considerable variation exists in the length of the cycle and its phases.

The menstrual cycle lasts twenty-eight days on average and comprises two phases: the follicular phase, which begins with the first day of menstrual bleeding and lasts until ovulation, and the luteal phase, which begins after ovulation and lasts until the last day before menstrual bleeding (Hawkins and Matzuk 2008). During the follicular phase, the main goal is to yield a mature follicle that will be able to release an egg in a process called ovulation. The follicle consists of an egg cell, called the oocyte, and two main types of supporting cells: granulosa cells and theca cells. At the beginning of each cycle, about fifteen to twenty primordial follicles start to grow and develop. Usually only one of them, the dominant follicle, will go through all stages of development and become a fully mature follicle. Primordial follicles, each with an oocyte inside, are produced through divisions of germ cells during fetal development; a human female is born with approximately 7 million of these follicles. After a woman is born, no new follicles will develop in her body, and the supply of follicles gets smaller with time.

The menstrual cycle needs the well-coordinated activity of several physiological partners: the hypothalamus, the pituitary, the ovary, and the endometrium or uterine lining. The hypothalamus, a part of the brain, secretes gonadotropin-releasing hormone (GnRH). The pituitary, an endocrine gland located at the base of the brain, secretes luteinizing hormone (LH) and follicle-stimulating hormone (FSH). Secretion of these two pituitary hormones is stimulated by GnRH. The ovary itself secretes estradiol, progesterone, and inhibins. The endometrium is lining of the uterus, which grows under the stimulation of ovarian hormones in preparation for possible implantation of a fertilized egg.

At the beginning of the cycle, levels of FSH increase and stimulate the maturation of primordial follicles, which develop from primordial germ cells. Each follicle begins to produce steroid hormones. The theca cells in a follicle secrete androgens (under stimulation by LH), which are converted by granulosa cells to estradiol. As already mentioned, the dominant

19

follicle usually grows faster and produces more estradiol than the other follicles. Estradiol exerts a suppressing effect on FSH production, and, in addition, the dominant follicle produces the hormone inhibin, which further lowers FSH production. As the levels of FSH diminish, all nondominant follicles without support of this hormone cannot continue further development and gradually die.

The dominant follicle continues to grow and secretes an increasing amount of estradiol. When the concentration of estradiol reaches a high level, it stimulates the pituitary to produce large amounts of LH (the LH surge). Under stimulation of LH, the follicle undergoes biochemical changes that cause it to rupture and expel the mature egg. The egg begins its journey toward the uterus via a duct called the fallopian tube, where fertilization occurs. The fertilized egg starts to develop immediately, and after a few days it is ready to implant in the uterus. The remains of the dominant follicle form the corpus luteum, which starts to secrete another steroid hormone—progesterone. Progesterone further prepares the lining of the uterus, which has already thickened under stimulation of estradiol, to be ready for implantation.

If fertilization has not occurred, the corpus luteum continues its production of hormones for about two weeks and then dies. Without support from progesterone, the uterus sheds its lining, which causes menstrual bleeding. Low levels of estradiol and progesterone during that time stimulate the hypothalamus and pituitary to produce their hormones, which in turn causes new recruitment of primary follicles, and a new cycle begins.

Hormone Levels and Chances of Conception

Two main reproductive steroid hormones produced during menstrual cycles, 17–β estradiol and progesterone, are involved in processes leading to ovulation, fertilization, and implantation of the fertilized egg. Their levels are important for the successful completion of these processes; as such, they are directly responsible for the establishment of pregnancy. Estradiol levels correlate with follicle size and with oocyte quality, and are related to the morphologic features and thickness of the endometrium (Eissa et al. 1986; Dickey et al. 1993; Lamb et al. 2011). Low levels of estradiol are associated with diminished penetrability of mucus in the female reproductive tract by sperm cells, which in turn contributes to reduced fertility (Roumen, Doesburg, and Rolland 1982; Katz, Slade, and Nakajima 1997).

Progesterone is essential for endometrial maturation; the higher levels this hormone reaches, the more complete the transformation of the endometrium in preparation for implantation of a fertilized egg (Santoro et al. 2000). During the first half of the cycle, the follicular phase, progesterone also seems to be important because it controls the proliferation and differentiation of the granulosa cells (Chaffkin, Luciano, and Peluso 1993).

It is understandable that a lack of progesterone production during the luteal phase would make implantation impossible, or that the absence of ovulation means that conception will certainly not occur. However, is more subtle variation in the levels of hormones important? If the cycle is ovulatory and estradiol and progesterone are produced, is the quantity of the produced hormone really that crucial? In the medical world, cycles are often merely classified as ovulatory or nonovulatory, and moderate variation in the hormone levels observable during ovulatory cycles is considered uninteresting and unimportant. We cannot know whether such variation is important until we show that cycles varying in levels of these hormones have corresponding variation in the probability of pregnancy. Only such a finding would prove that variation in hormone levels has functional importance.

In a group of white U.S. women who were attempting to become pregnant, women were more likely to conceive during menstrual cycles with higher levels of follicular estradiol (Lipson and Ellison 1996). In other words, these women could not become pregnant during some of their cycles. When hormones in saliva samples were measured, it became clear that nonconception cycles did indeed have lower levels of estradiol than cycles of the same women that ended in pregnancy. In the women who participated in this study, the average estradiol levels during the menstrual cycle were associated with a 12 percent probability of conception, while a 37 percent rise in estradiol levels led to a threefold increase in the probability of conception (to about 35 percent). In healthy Chinese women who were trying to conceive, cycles characterized by higher levels of estradiol resulted in higher rates of conception (Venners et al. 2006). Progesterone levels, especially during the midluteal phase, are also positively correlated with the chance of successful conception (Lu et al. 1999).

We may conclude that it is a well-established fact that, among women from the same population, lower levels of ovarian hormones lead to a lower chance of conception. It is less clear whether differences in the average levels of these hormones described for *different populations* can be interpreted in the same way. That is, if lower average levels of estradiol

21

and progesterone are recorded for a given population, does this lead to a generally lower probability of conception in that population? For example, are Nepalese women, in general, less likely to conceive than North American women? It is a legitimate question. The average progesterone levels in the midluteal part of the menstrual cycle are about 30 percent lower in women from Nepal in comparison with women from the United States (Ellison et al. 1993; Jasienska and Thune 2001a).

Some anthropologists have suggested that ovarian function may have different set points, depending on lifestyle conditions (Vitzthum 2001). Women living in chronically poor energetic conditions may have lower levels of ovarian hormones, but these levels are thought to be sufficient for conception to occur. For example, rural Bolivian women conceived during cycles when their mean progesterone levels were approximately 40 percent lower than those measured during conception cycles of urban U.S. women (Vitzthum, Spielvogel, and Thornburg 2004). Some reproductive ecologists have argued that women living in chronically poor energetic conditions are not expected to have levels of hormones as high as women having good energetic conditions (Ellison 1990; Lipson 2001). Lower levels of hormones during the cycle mean that it is more difficult for a woman to conceive; that is, the waiting time to conception may be longer, but conception is still possible.

It is important to note that comparisons of mean levels of ovarian hormones among populations or among individual women are difficult. One needs to make sure that women are of comparable age, are selected for participation in the study using similar criteria, and that differences in laboratory procedures do not contribute to variation in hormone levels. Further, hormonal concentrations measured in blood or serum are not directly comparable with values obtained from saliva or urine (Riad-Fahmy, Read, and Walker 1983; Riad-Fahmy et al. 1987; Ellison 1988; Shirtcliff et al. 2000). In consequence, reliable comparisons of hormone levels can be made, at present, only for very few populations (Ellison et al. 1993; Ellison 1994; Jasienska and Thune 2001a).

Why Do Women and Populations Differ in Levels of Hormones?

Human reproductive ecology is the area of science that attempts to quantify the magnitude of variation in reproductive function and identify its

22

causes. Research conducted in the last couple of decades in populations such as urban U.S. women, Polish and Nepalese farmers, African and South American hunter-gatherers, and Bangladeshi migrants to the United Kingdom has brought rich sets of data and has led to building theoretical frameworks aimed at explaining variation in reproductive physiology.

Especially significant contributions to human reproductive ecology have been those of Peter Ellison from Harvard University. Among many hypotheses and results, two aspects of his work are perhaps the most important. His training in zoology and evolutionary biology allowed him to look at human reproduction in an innovative way and provide evolutionary explanations of observed phenomena. For example, he argued that ovarian suppression is not a pathology but an evolved, adaptive response to environmental stresses. Furthermore, laboratory methods have been developed by Ellison's research team that have opened new opportunities. Measuring steroid hormones in salivary samples has truly opened a new world to those interested in human reproduction. Saliva samples are noninvasive to collect and can be stored without freezing. Thanks to this technique, hormones can be measured in nonclinical settings, from people in some of the most remote populations of the world. For the first time, knowledge about variation in ovarian function in women and testicular function in men among the world's populations has begun to emerge.

Quite a lot is known about the causes of variation in reproductive function among healthy women. Levels of hormones are influenced by many factors: genes, developmental conditions during fetal and childhood growth, and adult lifestyle (Figure 1.4). It is also well established that levels of reproductive hormones change with a woman's age. The lowest levels are observed in the years after menarche (i.e., reproductive maturation marked by first menstrual bleeding) and before menopause, and the highest levels are found in women between twenty-five and thirty-five years of age (Ellison 1994). It is important to emphasize, however, that even among women of the same age substantial variation in ovarian function may be present.

Ovarian function is sensitive to changes in lifestyle and responds with reproductive suppression, especially to factors such as energy expenditure or negative energy balance, when energy intake is lower than the energy used by the organism. Ovarian suppression is understood as any change in ovarian function that lowers the chance of pregnancy, and it is thought

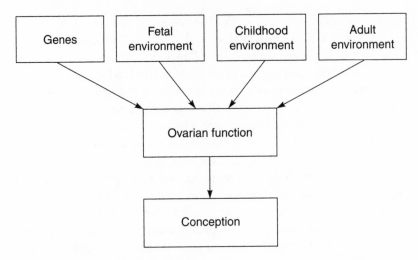

Figure 1.4. Determinants of variation in the level of ovarian steroid hormones.

to occur in a gradual fashion (Prior 1985; Ellison 1990). For example, a small reduction in body weight may cause low estrogen production in the menstrual cycle and thus a reduction in the chance of conception. A more serious weight loss may result in the absence of ovulation, and a dramatic weight loss may result in the total suppression of cycling (amenorrhea). The probability of pregnancy is, of course, reduced to zero when menstrual cycles are anovulatory or absent.

However, it should be stressed that variation in levels of hormones exists even among cycles that *are* ovulatory. As already mentioned, such variation is functionally important, because when cycles are ovulatory but are characterized by reduced levels of ovarian steroid hormones, the chances of successful conception are diminished (Eissa et al. 1986; Mc-Neely and Soules 1988; Dickey et al. 1993). Women who have ovulatory cycles with low levels of steroid hormones can still conceive during these cycles, but the probability that conception will occur is much lower than during the cycles with higher levels of steroid hormones. In environments with chronically low availability of energy, lower levels of hormones and the resulting lower chance of conception are beneficial for women. A longer waiting time to conception allows women to improve their own energetic status before they become pregnant again (Ellison 1990).

Hormones and Conditions during Childhood

Differences in ovarian function among adult women originate very early. Nutritional conditions as early as during fetal and childhood growth and development are important for the development of the reproductive function. Variations in such early experiences contribute to variation in the ovarian function of adult women. The importance of conditions during fetal development for ovarian function is discussed later on (Chapter 3), and other long-term consequences of fetal experiences are covered in Chapters 4 and 5.

Nutritional conditions during childhood are also important for reproductive function because they influence the age of sexual maturation (Ellison 1982, 1990; Vermeulen 1993; Belachew et al. 2011). Girls with poor nutritional status mature later than those with good nutritional status. Furthermore, in early maturers the levels of hormones in menstrual cycles are higher in comparison with late maturers for at least several years after menarche (Vihko and Apter 1984).

Recently, more direct evidence has pointed to the impact of childhood conditions on adult ovarian function. Alejandra Núñez-de la Mora, Gillian Bentley, and collaborators (2007) analyzed the levels of ovarian hormones in women who had experienced different childhood conditions. Bangladeshi women of reproductive age living in Bangladesh were compared with women who had migrated to the United Kingdom at various points of their lives. Significant differences among the different groups of Bangladeshi women were discovered. Women who had spent their childhood in Bangladesh had lower levels of progesterone than women who had been born in Bangladesh but migrated to the United Kingdom as children, or those who had been born in the United Kingdom. Among the migrants, the women who had migrated as young children (younger than eight years old) had significantly higher levels of progesterone than those who had migrated at a later age. They also had an earlier age at menarche and were significantly taller as adults.

For the women who had migrated as children before menarche, their age at migration predicted their age at menarche as well as their average levels of luteal progesterone: early migration was associated with earlier maturation and higher hormone levels in adulthood. Why do such differences occur in ovarian function? The girls who grew up in Bangladesh came from relatively affluent families and most likely did not experience nutritional stress while growing up, but they still were in poorer energetic

25

condition than girls who had grown up in the United Kingdom. The authors of the study suggested that the energetic stress resulted mostly from maintaining an elevated immune response to the high incidence of infectious and parasitic diseases present in Bangladesh but not in the United Kingdom. The necessity of allocating significant energy to immune function perhaps provided a signal for the developing individual that the environmental conditions were challenging and that if such challenges were to persist in future, less energy would be left to support reproduction.

However, the results of this study do not permit us to conclude that women with childhood energetic stress will have permanently reduced levels of progesterone as adults, or that they have an increased sensitivity of ovarian function to energetic stresses in adulthood. I will discuss (Chapter 3) the phenomenon of increased sensitivity of ovarian function as a consequence of energetic stress during fetal development. It would be interesting to learn from future studies whether energetic challenges during childhood change the subsequent sensitivity of ovarian function, or whether they cause permanent impairment in levels of ovarian hormones. Studies of women for whom not only childhood conditions but also fetal conditions are known would allow us to investigate the interactive effects of these two environments.

Energy Metabolism during Adult Life: Energy Intake and Physical Activity

Energy is the key to understanding ovarian function (Ellison 2003a, 2003b). Women who are in poor energetic condition during adult life often have reduced levels of steroid hormones. Poor energetic condition results from either high levels of energy expenditure or from low levels of energy intake. When energy expenditure is higher than energy intake, an individual experiences weight loss, a state that is called negative energy balance.

Intense exercise, usually resulting from participation in professional sports, and the resulting negative energy balance are associated with an increased incidence of menstrual disturbances and even a total lack of menstrual cycles (Feicht et al. 1978; Prior et al. 1982, 1992; Prior 1985; Broocks et al. 1990; Rosetta 1993, 2002; Rosetta et al. 1998; Redman and Loucks 2005). Some researchers have suggested that professional athletes represent a unique group of women, but it turns out that a similar disruption in ovarian function can be induced in previously untrained women who are subjected to experimentally imposed demanding regimes of aerobic exercise (Bullen et al. 1985).

Far more interesting are the findings that ovarian function is also affected in women engaged in recreational exercise. Such exercise almost never leads to a disruption of menstrual patterns, and cycles usually remain regular. However, changes in ovarian function do occur, although they are far less dramatic than those resulting from a professional level of exercise and usually include a lowering of the levels of ovarian steroids (Pirke et al. 1989; Broocks et al. 1990; Ellison 1990; De Souza et al. 1998; Jasienska and Ellison 1998, 2004; Jasienska, Ziomkiewicz, Thune et al. 2006; Stoddard et al. 2007). Exercise does not need to be very intense to induce some degree of ovarian suppression; for example, recreational joggers who run on average less than 12.5 miles (20 kilometers) per week have suppressed progesterone levels (Ellison and Lager 1985, 1986; Bledsoe, O'Rourke, and Ellison 1990).

Severe caloric restriction in women during the famine that occurred in the winter of 1944–1945 in German-occupied Netherlands resulted in a reduced ability to conceive (Painter, Roseboom, and Bleker 2005). We know this only from birth records. Ovarian hormones obviously were not measured, but it can be assumed that hormone levels were suppressed. More recent findings show that women who have lost about 15 percent of their initial body weight experience disturbances in the release of pituitary gonadotropins (hormones involved in stimulating production of ovarian steroid hormones) and consequent amenorrhea, a state characterized by the absence of menstrual cycles (Vigersky et al. 1977). Caloric restriction diets in young women frequently lead to menstrual disturbances and suppressed levels of ovarian steroid hormones (Pirke et al. 1985). Eating disorders such as anorexia nervosa and bulimia nervosa, which most frequently occur in girls or young women, are often associated with menstrual and hormonal problems (Becker et al. 1999).

Women do not need to be thin to develop ovarian suppression in response to weight loss. The loss of weight does not need to be substantial, either. Lower salivary progesterone levels have been observed in women of normal weight who have lost moderate amounts of weight through calorie-restriction dieting (Lager and Ellison 1990). In these women, the suppression of ovarian steroid levels was even more pronounced in the cycle following that in which the weight loss occurred. Suppression of ovarian function in response to a negative energy balance is stronger when weight loss occurs in combination with exercise (Bullen et al. 1985), or when it occurs in young women (Schweiger et al. 1989).

One may argue that exercise and dieting aimed at losing weight are novelties resulting from the demands of modern, urban life, and as such are "unnatural" for the organism. If so, then ovarian suppression represents an "unnatural" response to "unnatural" stimuli. Studies of women with more traditional, nonurban lifestyles were needed to answer the question of whether similar physiological responses occur in association with stimuli of a more "natural" type. These new studies undertaken by, among others, Peter Ellison, Gillian Bentley, Catherine Panter-Brick, Virginia Vitzthum, and Grazyna Jasienska, have shown that variation in ovarian function in association with variation in energetic factors is not restricted to cases of voluntary exercise or dieting in urban women. Ovarian suppression also occurs in women with traditional lifestyles in response to workload or seasonal food shortages. Farm women in rural Poland have profiles of salivary progesterone that vary with the intensity and duration of their workloads: heavier workloads mean lower hormone levels (Jasienska 1996; Jasienska and Ellison 1998, 2004). Women in Zaire and Nepal show seasonally suppressed levels of ovarian steroids when workload is heavy and energy balance is threatened (Ellison, Peacock, and Lager 1986; Panter-Brick, Lotstein, and Ellison 1993; Bentley, Harrigan, and Ellison 1998). In Zaire, this seasonal variation in ovarian function is probably the main cause of the observed seasonality in conceptions, as most births happen nine months after cycles characterized by the highest levels of hormones (Bailey et al. 1992).

It should be stressed that energy expenditure due to sports participation or occupational work may influence ovarian function independently of the sign of energy balance. Previously untrained women were subjected to eight weeks of an intense physical exercise regime (Bullen et al. 1985). While women in one group were losing weight during such training (i.e., they were in a state of negative energy balance), women randomly assigned to the second group were receiving a diet with enough additional calories to maintain their prestudy body weight. Even though the suppression of ovarian function was more pronounced in the weight-loss group, the women who did not lose weight while training also showed evidence of ovarian suppression.

Women who run on average 12.5 miles per week (about 20 kilometers) have suppressed ovarian function even though they maintain a stable body weight (Ellison and Lager 1986). Even though the runners' cycles were of similar length as those of nonactive women, the runners' cycles were characterized by a shorter luteal phase and lower levels of luteal progesterone. In women from Poland, the amount of habitual physical activity (as measured by the average, total daily energy expenditure) corresponded to their levels of estradiol (Jasienska, Ziomkiewicz, Thune et al. 2006). Women from the group with low activity had estradiol concentrations that were almost 30 percent higher than those of the women from the group with high activity.

Like exercise, occupational work, even when not causing negative energy balance, may also cause ovarian suppression (Jasienska 1996; Jasienska and Ellison 1998; Jasienska, Ziomkiewicz, Thune et al. 2006). Environmental settings that allow the study of the effects of intense physical work on ovarian function in women who are well-nourished are not easy to find. In most populations around the world, intense work coincides with rather poor nutritional status and frequent food shortages. However, well-nourished but still hardworking women still live in some areas of rural Poland.

The Mogielica Human Ecology Study Site in southern Poland was established to study the impact of a lifestyle associated with traditional, small-scale, labor-intensive agriculture on physiology and health. The study site includes five villages located in mountainous valleys where, owing to spatial localization and fragmentation of fields, agricultural labor is still mainly accomplished by hand and use of horses. Two main factors contribute to the uniqueness of this setting: (1) high energy expenditure, complemented by (2) high energy intake. People expend a lot of energy while working using traditional, old-fashioned agricultural methods, but at the same time they are able to maintain a relatively good nutritional status. Food shortages, often a salient feature of small-scale traditional food production, do not occur anymore in these villages. The diet is based on locally grown products and is supplemented by food purchased in stores and markets. Many people are able to sell grains, vegetables, fruit, milk, and meat, and others work for wages, generating cash for goods.

Farm work is very seasonal, and most women are not involved in work outdoors until May when they plant vegetable gardens. Most families own cows, and from mid-May these need to be walked to and from the pastures on a daily basis. Pastures are not always close to the stables where animals stay during the night, and often a distance of a few kilometers separate the stables from the grazing areas. Cows are milked two to three times per day, and this is traditionally a woman's job—most men do not even know how to do it. When women from the farm have plans to travel, they often arrange with a female neighbor to take care of the milking.

Work intensity increases in late June when grass for hay is cut—this is a man's task, often done with a scythe. Cut grass remains outside for a few days and needs to be turned daily, and after that needs to be arranged in haystacks. Both men and women know how to perform these tasks. Hay continues to dry outside for a week or so. Dry hay is transported to the farm and stored as winter feed for cows and horses. The harvest season starts in the second half of July, and haying may still continue during this month. Usually every farm grows several types of cereal, including wheat, rye, barley, and oats. Cereals, most often cut with a scythe, are immediately arranged into grain stacks, and left to dry. In July, black and red currants, gooseberries, and raspberries are picked from gardens, and wild mushrooms and blueberries are gathered in the woods, often in substantial quantities. Women bring them home in buckets, make fruit preserves, and store them in jars for the winter. In August, the harvest continues, and cereals are transported and subsequently threshed with the use of threshing machines. This requires substantial work effort because cereals must be loaded into the machine, and later grains and remaining thatch must be packed and transported to storage.

During August, the second hay of the season is collected, transported and stored. Soon after the cereals are cleared from the fields, the ground is plowed with horses in preparation for the next planting season. In September, potatoes are harvested, and "winter crops" of wheat and rye are sown (to be harvested in July and August of the next year). In October and November, men cut and transport wood, which will be used for heating and cooking throughout the year. During the winter months, women spin wool, knit sweaters, and sew clothing and linens, although these activities are mostly done by the older generation, and such traditions are slowly disappearing.

Although agricultural fieldwork is very seasonal, women have daily duties of tending children and animals all year round. They also do most of the house cleaning, cooking, and laundry. Many women still bake bread and make pasta, butter, and cheeses. Time spent on these activities does not vary substantially across the year, although at harvest time less elaborate cooking is done compared with the rest of the year (Jasienska 1996). Substantial seasonal differences in mean daily energy expenditures observed for women are due mainly to the seasonal nature of the fieldwork. The sexual division of labor is quite pronounced, and women are almost never involved in work requiring the use of horsepower (e.g., plowing, sowing, and fertilizing with manure) or a scythe. Men are generally not involved in housework or many forms of animal care, and only occasionally help with childcare. Children are drawn relatively early into many types of activities, helping parents with the harvest, haying, animal care, housecleaning, shopping, and taking care of younger siblings.

The agricultural work is traditional, but many other aspects of their lives are quite modern. There are television sets, refrigerators, and washing machines in every house. When I was doing fieldwork for my doctoral thesis on the relationship between physical work and ovarian function, I visited women participating in the project in their homes to conduct dietary and physical activity interviews and to take anthropometric measurements. I quickly learned that some times of the day are better than others. In the 1990s, nearly every woman in these villages followed a Brazilian soap opera—the first soap opera ever shown on Polish television. Before I discovered this, I would visit them during this special hour and was politely invited to sit down and watch entire episodes with all women of the house, from granddaughters to grandmothers; only after it had ended was I able to get the attention of my study participants. Unsurprisingly, a subpopulation of Polish girls born in that period have South American first names.

Women from the Mogielica Human Ecology Study Site collected saliva samples daily for six months, and had their nutritional status assessed by anthropometric measurements and their energy intake and expenditure evaluated by twenty-four-hour recall questionnaires. This long-lasting sample collection was necessary to test a hypothesis that seasonal changes in lifestyle have an impact on ovarian function. The completeness of the saliva sample collection was exceptional (partly owing to the help of the children who liked the idea of their mothers "spitting" and would remind them to do it). Women's salivary progesterone levels were reduced by almost 25 percent during the months of intense harvest-related activities. In these months, a linear relationship between the levels of mean total energy expenditure and progesterone levels was observed: women with the highest energy expenditure had the most pronounced ovarian suppression.

Not only were these women in good nutritional status with a mean body mass index (BMI, calculated as body weight in kilograms divided by body height in meters squared) of $24.4 \, \text{kg/m}^2$ and a mean body fat percentage of 27.5, but they also did not fall into the state of negative energy

31

balance as a result of intense work. Statistical models indicated that neither body weight, nor fat percentage, nor energy balance had any relationship with their progesterone levels—only energy expenditure was important. These results clearly suggest that physical work may lead to a suppression of ovarian function even in women who have good nutritional status, have sufficient fat reserves, and do not experience negative energy balance.

The results reviewed here clearly indicate the importance of energetic factors for ovarian function. Ovarian suppression is most likely caused by *changes* in the availability of metabolic energy rather than the absolute amount of energy. Changes in the availability of metabolic energy can result from an increase in energy expenditure or a reduction in energy intake. These changes send a message to the woman's body that the availability of energy has declined or that an increased demand for energy from physiological processes other than reproduction has arisen.

Does Body Fat Improve or Worsen Fecundity?

Reproduction requires energy. In humans, substantial amounts of energy can be stored as fat. An obvious conclusion can be drawn based on these two facts: women with larger fat stores can reproduce better. Alas, this is a case in which H. L. Mencken's admonition applies: "There is always an easy solution to every human problem—neat, plausible and wrong."

It is true that women with very low body fat levels, especially when resulting from anorexia nervosa, often experience menstrual and hormonal disturbances, and thus are less likely to conceive. This situation seems to support the "body fat–reproduction" hypothesis. But similar menstrual and hormonal problems are often present in women with very high body fat levels. Although several studies have described a positive or an inverse U-shape relationship between BMI or body fat percentage and estradiol levels in women (Bruning et al. 1992; Barnett et al. 2002; Furberg et al. 2005; Ziomkiewicz 2006; Ziomkiewicz et al. 2008), none of these were able to control for a very important factor: energy balance. It is likely that variation in BMI or body fat is correlated with variation in energy balance.

Many women in urbanized societies may be able to achieve low BMI or fat percentages by striving to maintain low-energy diets, or actively exercising, or both. Therefore, these women are more likely, on the average, to be in a state of negative energy balance or to have high levels of energy expenditure. Both factors have a well-established suppressive effect on ovarian function. Negative energy balance or high energy expenditure are the real causes of low steroid levels in women with low levels of fat, while fat stores by themselves may have nothing to do with their ovarian function. Studies that carefully assess all these factors are necessary to prove that low ovarian function in women with low body fat is really related to body fat reserves and not to negative energy balance or exercise, but so far no convincing evidence supports the direct importance of body fat for ovarian function.

In fact, an excess of body fat seems to be detrimental to fertility. Women who are obese or overweight have a lower chance of conceiving and suffer more complications during pregnancy than women of a healthy body weight (ESHRE Capri Workshop Group 2006; Homan, Davies, and Norman 2007). Most data about a relationship between body weight and fecundity come from studies of women who have problems becoming pregnant and who are patients of fertility clinics. For example, a study that examined infertility patients from the United States and Canada found that overweight and obese women (with a BMI of at least 27) were three times less likely to conceive than women with a BMI between 20 and 25 (Grodstein, Goldman, and Cramer 1994).

Obesity decreases the chances of pregnancy even among women who are not seeking medical help for fertility problems. Obese British women reported more menstrual problems and were less likely to conceive than women with lower body weights (Lake, Power, and Cole 1997). Once they conceived, they often experienced complications such as hypertension. In the United States, the risk of infertility increased from 1.1 among women with a BMI between 22.0 to 23.9 to 2.7 in women with a BMI of 32 or greater (Rich-Edwards et al. 1994). In fact, women in all categories of BMI above 23.9 had an elevated risk of infertility.

Body fat, a metabolically active tissue, converts androgens produced by the adrenal gland to estrogen. In obese women, high levels of these estrogens interfere with the production of estradiol by the ovary; as a result, circulating levels of estradiol are often low (De Pergola et al. 2006). Obese women may also have high levels of testosterone, with potentially detrimental effects on follicles, oocytes, and the endometrium (Gosman, Katcher, and Legro 2006).

Improved menstrual regularity, higher rates of ovulation, higher levels of hormones, and a resulting higher chance of conception have been observed in obese women who lost a modest amount of weight; even a loss of less than 10 percent seems to be effective (Falsetti et al. 1992; Clark et al. 1995; Galletly et al. 1996; Norman and Clark 1998). Weight loss is, therefore, recommended for obese women who have been unable to conceive. Some authors have even suggested that weight loss programs should be offered to women before any other treatments for infertility are implemented (Norman et al. 2004).

Hourglass Figure and Hormone Levels

Negative energy balance or a rise in energy expenditure due to physical activity clearly causes a reduction in levels of reproductive steroid hormones, although the amount of energy that is stored as fat is probably not very important for the fecundity of women with a normal body weight. When one excludes women with very low or very high body fat, having less or more body fat does not affect the levels of reproductive hormones. What does matter, however, is the pattern of fat distribution: whether fat is deposited on the hips or waist, and the breasts.

The 1902 Sears catalog advertised the Princess Bust Developer and Bust Cream or Food as "the only treatment that will actually, permanently . . . cause [women's breasts] to fill out to nature's full proportions, give that swelling, rounded, firm white bosom, that queenly bearing, so attractive to the opposite sex" (Wolf 2001, 24). An advertisement for another product—the Madame Mozelle Compound Bust Developing Treatment said, "It is a matter of statistics that the great majority of neglected wives and old maids are . . . flat-chested . . . Don't be one of the gray cohorts of hopeless femininty [sic] when it is so easy to attain that physical perfection that will turn you into an enchanting creature of many allurements" (Wolf 2001, 24).

The fertility rate—the number of children born, on average, to a woman during her entire reproductive span—in the United States decreased from 7.04 in 1800 to 4.24 in 1880, and further to 3.56 in 1900 (Wolf 2001, 23). Some have claimed that, in the nineteenth century, marriages became less about reproduction and more about having romantic relationships, and at that time women's breasts changed their function from just producing milk for children to having sexual significance for their husbands (Wolf 2001, 23). But is the obsession with women's breasts just a cultural phenomenon? Or do breasts have cultural importance *because* they are somehow related to female reproductive potential?

In our study of Polish women, we attempted to address this question (Jasienska et al. 2004). Women between twenty-four and thirty-seven years of age collected saliva samples during an entire menstrual cycle. These samples were used to measure levels of estradiol and progesterone. Just before the women began collecting the samples, we measured their breast size, and waist and hip circumferences. For each woman, her breast size was calculated as the ratio between two measurements: the circumference of the breast at the widest part and the circumference directly under the breasts. What is left is a measure of breast size that is independent of how heavyset or thin a woman is.

Relative size of the waist is another trait used to characterize female body shape. Relative waist size is assessed by measuring the waist and hips, then calculating the waist-to-hip ratio (WHR). According to human evolutionary psychology, physical characteristics such as breast size and especially WHR function as features used by human males to assess female attractiveness. Males, at least in industrialized countries, pay attention to these features because (evolutionary psychologists say) they serve as cues to fecundity and health.

Such hypotheses are difficult to prove because only a few studies have reported a significant relationship between low WHR (narrow waist) or large breast size and increased fecundity. In addition, most of these studies had methodological problems: women with a high WHR (large waistline but relatively narrow hips) were also obese (Evans et al. 1983; Moran et al. 1999) or were patients of fertility clinics (Zaadstra et al. 1993). Low fertility in such women may be a direct effect of obesity, not of high WHR. In our Polish study, the relationship between body shape and fecundity was investigated in women who were healthy, had no reported fertility problems, and were not obese (Jasienska et al. 2004).

Women with a higher breast-to-underbreast ratio (large breasts) and women with a relatively low WHR (narrow waists) had significantly higher fecundity as assessed by their levels of estradiol (Figure 1.5) and progesterone. Even more interesting was the finding that women who had both a narrow waist and large breasts had 26 percent higher mean estradiol and 37 percent higher mean midcycle estradiol levels than women from three groups with other combinations of body shape variables (i.e., low WHR with small breasts, or high WHR with either large or small breasts). In this study, breast sizes and WHR were large or small in a relative sense; that is, the mean values of a study group were used as criteria for categorizing women into groups with large or small sizes. It means that the relationship between body shape and reproductive potential can be detected even among "average" women. These results show that fat distribution, which to a significant extent determines shape of a female

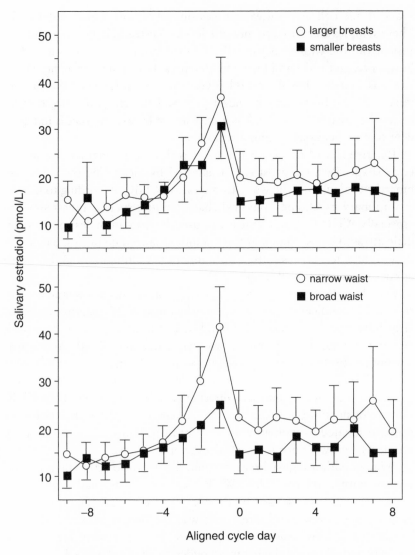

Figure 1.5. Profiles (95 percent confidence intervals) of salivary estradiol in groups of women with larger and smaller breasts and narrow and broad waists. Daily estradiol values were aligned for each cycle on the day of midcycle estradiol drop (day 0).

body, may be an important factor contributing to fecundity. No wonder that human males in many cultures find small waists and large breasts attractive.

Interactions between Lifestyle Factors and between Genes and Lifestyle

We reviewed some evidence that, in healthy women of comparable age, variation in hormone levels can be to a large extent explained by variation in energetic factors, which are, in turn, related to differences in the environmental conditions these women experience at different stages of life.

As will be discussed in Chapter 2, genetic variation is also important when it comes to influencing levels of ovarian hormones. Women who have different genotypes with respect to the genes encoding enzymes involved in the synthesis of steroid hormones differ in their levels of these hormones. Such genotypic differences are often found, but truly interesting questions emerge when one addresses the ubiquitous effects of so-called genotype-environment interactions. Such interactions exist when the magnitude of the observable differences depends on particular environmental conditions in which the genotypes are studied. Do women with genetically low hormone levels have responses to environmental energetic factors that are comparable with women who have genetically high hormone levels? Would physical exercise of exactly the same duration and intensity have the same effect on women of different genotypes? No studies so far have addressed this question.

If the suppression of reproductive function is an adaptive response to a low availability of metabolic energy, should similar responses be observed in all women, regardless of whether they are genetically low- or genetically high-hormone producers? Furthermore, as will be suggested in this book, genetic variation in hormone levels may be a relatively new evolutionary phenomenon, appearing at about the time of the origin of agriculture (i.e., about 10,000 years ago). If so, then it is very likely that not enough time may have elapsed for different genotype-specific responses to various environmental challenges to evolve.

The importance of the interactive effects of different lifestyle (or environmental) factors is another interesting issue to be addressed. Lifestyle factors during adulthood will also interact. Physical activity and weight loss independently cause ovarian suppression, but if they occur together the suppression is more pronounced. In the study by Bullen and coauthors

(1985) described earlier that looked at the effects of intense exercise regime on reproductive function, women who both exercised and lost weight had more severe ovarian suppression than did the women for whom exercise did not lead to weight changes.

Even more interesting are the interactive effect of factors operating at different life stages. Developmental fetal conditions interact with physical activity in adulthood, as will be described in Chapter 3. Childhood conditions also seem to influence ovarian function. However, we do not know if girls growing up in different energetic conditions react to the same level of energy expenditure in adulthood with ovarian suppression of similar severity or to the same amount of weight loss in adulthood, or whether childhood conditions change the sensitivity of adult ovarian response.

Answers to such questions are very important for the area of disease prevention for women, and they are especially crucial for prevention of hormone-dependent cancers, including breast cancer. The risk of such cancers can be reduced thanks to changes in lifestyle that have the power to change the levels of reproductive steroids produced during a woman's lifetime. If adult women indeed have varying sensitivities of ovarian function to exercise or weight loss depending on their fetal or childhood physiological experience, then "prescriptions" for lifestyle changes should take such differences in early environments into account. In practice, this may mean that different women may need different amounts of exercise or weight loss to achieve similar changes in their levels of reproductive steroid hormones and experience similar reductions in cancer risk.

Ovarian Suppression as an Evolutionarily Adaptive Phenomenon

Suppressed ovarian function reduces or prevents the occurrence of conception, so it would appear to be a pathological phenomenon. However, Peter Ellison (1990, 2003a, 2003b) has proposed that the physiological responses of the reproductive system of contemporary women to energetic factors are not pathologies but important features of human biology that developed 100,000 years ago in the Paleolithic era during the evolution of modern humans, long before the introduction of agriculture. Human physiology has changed very little since that time (Eaton, Konner, and Shostak 1988; Eaton, Strassman et al. 2002), and energetic stresses have been and still are salient features of life in many traditional populations (Roberts et al. 1982; Lawrence and Whitehead 1988; Panter-Brick 1993; Adams 1995; Benefice, Simondon, and Malina 1996; Sellen 2000).

During my classes with advanced, sixth-year medical school students, I always ask why individual women differ in their levels of ovarian hormones. After having explained that I am not interested in the influence of various diseases and disorders but in the causes of variation in healthy women, the standard, and the only, answer I receive is "fat." These future physicians are convinced that only body fat is important for the reproductive function in women.

In a set of highly influential papers, Rose Frisch (Frisch and McArthur 1974; Frisch 1985, 1987, 1990) proposed that, due to the high energy costs of pregnancy and lactation (which I will discuss further in Chapter 6), ovarian function should respond in an "on/off" fashion to body fat stores. A woman with insufficient body fat to support the energy costs of pregnancy and lactation should be unable to conceive. However, although fat stores may indeed be crucial for partially covering the energy costs of milk production (McNamara 1995), such stores are vastly insufficient to cover the energy costs of both pregnancy and lactation. In addition, women in developing countries often have very low fat reserves (Lawrence, Coward et al. 1987; Little, Leslie, and Campbell 1992; Panter-Brick 1996). It is also likely that during the human evolutionary past women did not have large fat stores. Australian Aborigines, for example, when living as hunter-gatherers, had BMIs well below 20 (O'Dea 1991), which indicates very low fat levels.

The diet of our ancestors had rather low energy density—the amount of energy per amount of food, usually expressed in kilocalories/gram or kilocalories/milliliter (fats and oils in purified form have the highest energy densities). Additionally, substantial energy had to be expended to obtain food. It is true that accumulated body fat is used by the maternal organism to help with the energy costs of lactation when her energy intake is not high enough. Unfortunately, body fat reserves, even when they are as substantial as they are in well-nourished Western women, are sufficient to cover only half of the costs of eleven months of lactation (Prentice and Prentice 1990; Lunn 1994). During the time before the beginning of agriculture, women probably practiced much longer breast-feeding, two or even three years per child. It is therefore unlikely that such a tight dependence of reproduction on body fat stores in human females (as proposed by Rose Frisch) would have been promoted by natural selection.

Reproductive suppression occurs when environmental conditions temporally deteriorate and women experience increased energy expenditure and often negative energy balance. Reproductive suppression serves to protect the maternal condition and to optimize a woman's lifetime reproductive output (Ellison 2003a). It lengthens the interbirth intervals and, by providing the comfort of metabolizing for one, allows women to improve their own nutritional status before the next energy drain. All acquired energy goes only to the maternal organism, without sharing it with a fetus or nursling.

Longer interbirth intervals mean that a woman will have fewer children. How can having fewer children be adaptive? In humans, important trade-offs appear to exist between the number and the quality of offspring (Strassmann and Mace 2008). Beverly Strassmann and Brenda Gillespie (2002) have found among Dogon women in Mali (West Africa) that very high fertility leads to lower reproductive success, measured as the number of children who survive to the age of ten. Chances of surviving the first five years of life were lower for children who had more siblings (Strassmann 1997b), clearly suggesting that in large families each child was less likely to receive adequate parental investment, which in turn led to poorer condition of the children and worsened survival.

Negative energy balance is a transient state, often occurring seasonally and quickly disappearing when food availability improves or workload becomes lighter. Ellison's hypothesis elegantly explains why reproductive suppression is adaptive in women with negative energy balance. However, the case of suppression occurring in response to high levels of physical activity in women who have a neutral or positive energy balance (i.e., are not losing or may be gaining body weight) requires additional explanation.

I propose a new explanation. The "constrained down-regulation" hypothesis suggests that intense workload compromises a woman's ability to allocate sufficient energy to reproduction (Jasienska 2001, 2003). Women who, as a result of an increase in physical activity, remain in a state of high energy flux (high energy expenditure, but compensated by high energy intake) may have an impaired ability to down-regulate their own metabolism when faced with the increasing energy needs of pregnancy and lactation. This creates a problem because the ability to lower basal metabolism is an important strategy to allow the mother to allocate more energy to the growing fetus (Poppitt et al. 1993, 1994; Sjodin et al. 1996). In the rural Gambia, women with poor nutritional status reduce their basal metabolism during pregnancy or lactation (Poppitt et al. 1993). When basal metabolism is reduced, some physiological or metabolic functions of the maternal organism do not get as much energy as they require, but additional energy can be allocated to the developing child.

Women who are experiencing high levels of energy expenditure may have a problem with reducing their own basal metabolism. In fact, in women with high levels of energy expenditure, basal metabolism tends to go up (Sjodin et al. 1996). It is possible then that when hard-working women have an elevated basal metabolism, their ability to manipulate it to redirect energy to reproductive processes is constrained. Temporary suppression of ovarian function may then be adaptive even in individuals who are still sustaining positive energy balance when their levels of physical activity are high.

Are Hormone Levels in Western Women Much Too High?

Medical science has usually assumed that urban women from industrialized countries have a physiology operating at optimal levels, so high levels of ovarian hormones in menstrual cycles are considered the physiological norm. The fact that much lower levels, such as those found in women from non-Western populations, are also observed is seldom discussed in the medical literature. The majority of medical doctors familiar with the reproductive physiology of Western women would most likely consider the hormone levels of women from other populations as being pathologically low. However, as pointed out by Peter Ellison (2003a), the hormone levels of non-Western women are not abnormally low—it is the hormone levels of Western women that are abnormally high.

Abundant energy availability during fetal and childhood development and during adult life contribute to high levels of hormones in menstrual cycles. Such good energetic conditions and correspondingly high levels of ovarian hormones were unlikely features during human evolution. When women were on a relatively tight energy budget, less energy went to the developing fetus. Growth during childhood was slower, and energy availability during adult life fluctuated, with periods of energy shortage occurring from time to time. All these factors contributed to setting the concentrations of ovarian hormones in our human female ancestors at much lower levels.

In Western populations, in addition to having high levels of hormones during most of their cycles, women have a high number of cycles during their lives (Strassmann 1997a; Eaton and Eaton III 1999). Early age at menarche and late age at menopause expand the range of years during which cycles occur. A few pregnancies suppress cycles for a relatively short time. If a woman has only a few pregnancies, she has a relatively short time of total lifetime breastfeeding. Even while breastfeeding, a woman resumes her cycles early, probably because nursing episodes are not very frequent and because she has, in general, good nutritional status (Valeggia and Ellison 2004).

Determinants or Merely Correlates of Hormone Levels?

Although some of the factors described here clearly causally influence levels of hormones, others merely show correlation with hormone levels. Results of cross-sectional, observational studies allow us only to conclude that two variables are correlated. For example, WHR and estrogen levels are correlated, but no claim of a causal relationship can be made. To complicate things further, physiological evidence suggests that estrogen impacts the pattern of fat distribution, so women with higher estrogen levels may have higher fat deposition in the hip region, lower fat deposition in the waist region, or both. At the same time, body fat may influence estrogen levels, and women with high abdominal fat may have reduced levels of this hormone. The causality here is of a complex, bidirectional nature.

Exercise or work-related physical activity seems to be, in contrast, *causing* changes in estrogen levels, not just correlating with them. It is unlikely (although in principle not impossible) that this relationship could run in the opposite direction, but that would mean that reduction in estrogen levels would result in an increase in physical activity levels. This relationship also has been investigated in experimental studies, with other potentially confounding factors carefully controlled (for example, Bullen et al. 1985). The long-term effects of energetic conditions experienced during fetal or childhood development can also be understood as causal, not just correlational. Although the physiological mechanisms involved are unknown, poor fetal conditions (as we will discuss later) seem to change the sensitivity of ovarian response, thus preparing the individual for similarly poor conditions expected in adulthood.

We have identified and discussed many factors responsible for the presence of substantial variation among women in their levels of reproductive hormones. Most of these factors are tightly linked to the availability of metabolic energy at various stages of a woman's life. The ability to reduce levels of reproductive hormones, which leads to a lower probability of conception in a given cycle, is an adaptive response of female reproductive physiology. One may hypothesize that this ability to temporarily suppress reproduction may increase the lifetime reproductive output of the mother, because longer intervals between births may help to improve the condition of both the mother and her children.

Knowing the factors that affect the level of reproductive hormones is also crucial for disease prevention. Relatively simple lifestyle modifications may be reflected in changes in the level of hormones, and such changes impact the risk of infertility, hormone-dependent cancers, osteoporosis, and other diseases.

In this chapter, I have mainly talked about the determinants of reproductive hormones that are part of one's lifestyle. In Chapter 2, I discuss the genetic dimension of the problem: genetic variation among women may also contribute to variation in the level of hormones.

High levels of estrogen are crucial for successful reproduction in women, but the estrogen concentrations present in the body are influenced by many factors, mostly related to metabolic energy. We know now that temporal suppression of estrogen levels in response to unfavorable environmental conditions may be considered adaptive. However, some women have permanently low levels of estrogen due to their DNA.

Alleles (variants of genes) that encode high levels of estrogen ought to be promoted by natural selection and be present in most women. A heightened ability to conceive is, after all, an important component of reproductive success. But, surprisingly, this is not the case: alleles coding for high estrogen levels are not universally predominant, and in all the studied populations they occur less frequently than alleles coding for low estrogen levels. Chapter 2 attempts to explain the paradox of the relative rarity of alleles coding for high levels of steroid hormones (which I call "high-level alleles"). I will argue that it is likely that high-level alleles became beneficial during human evolution only after the origin of agriculture, when humans began to consume large amounts of foods potent enough to reduce the levels of endogenous (i.e., produced by the body) steroid hormones.

2

Agriculture and Selection for High Levels of Estrogen

The term "genetic variation" (or genetic polymorphism) means that all individuals in the population do not have identical genetic makeup with respect to a particular gene. Why do we observe genetic variation in the genes responsible for the levels of reproductive hormones? Because high levels of hormones are important determinants of female fertility, one would expect a strong selective pressure for alleles encoding high levels of reproductive hormones. These alleles should be favored by natural selection and by now should be prevalent in all populations. Instead, in all studied populations there is considerable polymorphism in the genes involved in steroid production and metabolism. Why are low-level alleles present in modern populations at all?

I am going to discuss several steps that might have led from the origin of agriculture to the presently observed frequency of alleles responsible for levels of estrogen in women:

Step 1: The introduction of agriculture increases human consumption of plant chemicals called phytoestrogens.

Step 2: An agriculturally based diet provides a favorable, high-carbohydrate intestinal environment, rich in bacteria, that aids phytoestrogen synthesis from dietary precursors, consequently increasing the levels of high-potency phytoestrogens in the body.

Step 3: Phytoestrogens show high affinity for estrogen receptors and can also interfere with the synthesis of endogenous estrogens.

Step 4: Phytoestrogen consumption reduces the levels of endogenous steroid hormones.

Step 5: Reduced levels of ovarian steroids lead to reduced fecundity.

Step 6: Populations with higher phytoestrogen consumption and a longer history of such consumption should have a higher frequency of high-level alleles.

Genes Involved in Biosynthesis of Steroid Hormones

Many genes have the potential to influence female reproductive physiology. Crucial for circulating levels of hormones are the genes involved in the steroid metabolic pathway, which code for enzymes involved in steroid production. Genetic variants of *CYP17, CYP19, CYP1A1,* and *CYP1B1* (cytochrome genes) can potentially influence the levels of circulating steroid hormones. Of these, *CYP17* encodes cytochrome P450c17alpha, which mediates the activities of two enzymes: 17alpha-hydroxylase and 17,20-lyase. Both enzymes are involved in the biosynthesis of estrogen (Figure 2.1). In women, *CYP17* is expressed in the ovary, corpus luteum, adrenal gland, and adipose tissue (Small et al. 2005). A single-nucleotide polymorphism (i.e., a small change in DNA sequence, when various alleles of the same gene differ in having one different nucleotide) in one part of the *CYP17* sequence is relatively common, and the A2 allele is thought to increase transcription rates. The process of transcription is the synthesis of RNA based on information provided by the DNA. The RNA, in turn, is involved in the synthesis of proteins, including all enzymes.

In other words, it is thought that having a simple mutation in *CYP17* (i.e., having the A2 allele as part of the *CYP17* genotype) leads to faster production of enzymes that are needed for the synthesis of estrogens. *CYP17* is the most intensely studied of all genes involved in steroid metabolism, and its polymorphism is related to variation in levels of steroid hormones and also to incidence of hormone-dependent cancers. Because most data in the literature on issues relevant to this discussion concern the polymorphism of *CYP17,* I am going to concentrate on this gene when discussing genetic variation.

Although genes are responsible for influencing levels of all steroid hormones, most research so far has focused on estradiol levels, so this steroid will be the main hormone discussed here. Similar arguments may also be made when attempting to understand variation in other genes involved in steroid metabolism, especially *CYP19* encoding the enzyme aromatase, and also in genes that are encoding steroid receptors.

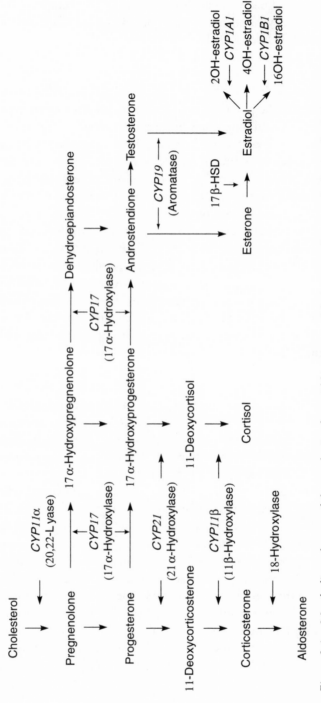

Figure 2.1. Metabolic pathways of the synthesis of steroid hormones with the genes that encode the involved enzymes.

From Agriculture to High Estrogen Levels: A Hypothesis

Why is there polymorphism among alleles coding for reproductive steroids, even though women having the high-level estrogen alleles would have a clear selective advantage over women who have, as part of their genotype, the alleles coding for lower steroid levels? One possible scenario is that high-level estrogen alleles were not selected because the women who had them had lower fitness than other women. How is that possible? Although estrogens are, in general, beneficial for all fertility and health-related features in women, their high levels are also responsible for an increased risk of hormone-dependent cancers, including breast cancer. Therefore, women with high levels of estradiol would have the advantage of having higher potential fertility but at the same time the disadvantage of a higher risk of dying from reproductive cancers.

A lower risk for reproductive cancers cannot, however, explain weak selection (or lack of selection) for high-level estrogen alleles. Reproductive cancers are relatively rare in women before menopause, and ovarian and endometrial cancers are rare in general. Breast cancer, the most common of all hormone-dependent cancers, most frequently occurs in older women, those in their sixth and seventh decades. Therefore, the advantage of having increased fertility as a result of having high-level estrogen alleles is unlikely to be canceled out by the disadvantage of higher mortality in postreproductive years. On the other hand, women in their sixth and seventh decades of life are grandmothers; in some societies such as the Hadza from Tanzania or in the rural Gambia, they have a very important influence on the survival of their grandchildren (Hawkes et al. 1998; Sear, Mace, and McGregor 2000). The fitness of the individual is the outcome of not only that individual's direct reproductive activity but also that individual's helping his or her relatives survive or reproduce. Fitness that takes into account such help—an indirect component called *inclusive fitness*.

However, even if being a grandmother is important for a woman's inclusive fitness, there still remains at least one more very serious problem with this hypothesis. Hormone-dependent cancers were probably very rare in human evolutionary history (Eaton and Eaton III 1999). Women were affected by few risk factors that would increase the likelihood of these cancers. They had a late age at menarche, a short period from menarche to the first birth, a relatively large number of children, and a long lifetime duration of total breastfeeding. All these factors meant that women foragers had far fewer menstrual cycles during which estrogen and progesterone were produced than do women today. Having fewer cycles resulted in having low lifetime levels of ovarian steroids. We know that low levels of estrogens and progesterone decrease the risk of having breast or uterine cancer.

How many menstrual cycles does a woman have during her life? There is a lot of variation among populations—the answer depends on her reproductive pattern and lifestyle. The Dogon women of Mali, who do not use contraception, provide a unique setting in which to address this question (Strassmann 1997b). Among the Dogon, who are millet-growing farmers, menstruating women are required to spend a night at a special menstrual hut, which means all menstruations can be recorded. Beverly Strassmann observed fifty-eight Dogon women and collected information on 477 cycles. She concluded that, on average, a woman in this society has only 110 menstruations during her lifetime. Women begin to menstruate when they are sixteen years old and finish having cycles when they are about fifty years old. During these thirty-four "cycling" years, over 440 cycles would be theoretically possible, given that during each year a woman can have thirteen 28-day-long cycles. But the Dogon women have high fertility of about 8.6 live births per woman. Lack of cycles during pregnancy and postpartum amenorrhea contribute to the low number of cycles during the entire lifetime in these women.

Eaton and coauthors (1994) estimated that contemporary hunter-gatherer women, a model for human ancestral great-grandmothers, have about 160 ovulatory cycles per lifetime. In contrast, a modern American woman has 450 ovulatory cycles. In addition, this number of ovulatory cycles is most likely an overestimate for hunter-gatherer women because many of their cycles have suppressed hormonal production during times when the women are in poor nutritional status. Thus, in contrast with American women who produce high levels of hormones during most of their 450 cycles, many of the 160 cycles of hunter-gatherers have suppressed hormonal levels, which contributes to their low lifetime exposure to ovarian steroids. Eaton and coauthors estimated the relative risk of cancers for contemporary American women to be twenty-four times higher for ovarian cancer, 114 times higher for breast cancer, and 240 times higher for endometrial cancer than for the hunter-gatherer women of the Paleolithic period.

As reproductive cancers were unlikely to be a major selective force against high hormonal levels, why do estrogen-producing genes still show polymorphisms in contemporary populations? There clearly was enough

PALEOLITHIC HUNTER-GATHERERS

```
                    ┌──────────────────┐
                    │   Low dietary    │
                    │  phytoestrogens  │
                    └──────────────────┘
    ┌───────────┐          │
    │ no effect │ ✕        │
    └───────────┘          ▼
                    ┌──────────────────┐     ┌──────────────────┐
                    │  Estrogen levels │     │ Low-level alleles│
                    └──────────────────┘     │ = ancestral state│
                       ↙          ↘          └──────────────────┘
          ┌──────────────┐   ┌──────────────┐
          │  Fecundity   │   │    Health    │
          │    HIGH      │   │    GOOD      │
          └──────────────┘   └──────────────┘
```

AGRICULTURAL POPULATIONS

```
                    ┌──────────────────┐
                    │   High dietary   │
                    │  phytoestrogens  │
                    └──────────────────┘
┌───────────────────┐     │
│ suppressive effect │     │
└───────────────────┘     ▼
                    ┌──────────────────┐     ┌──────────────────┐
                    │  Estrogen levels │     │ Selection for high-│
                    └──────────────────┘     │   level alleles    │
                       ↙          ↘          └──────────────────┘
          ┌──────────────┐   ┌──────────────┐
          │  Fecundity   │   │    Health    │
          │    LOW       │   │    POOR      │
          └──────────────┘   └──────────────┘
```

Figure 2.2. A schematic model shows that selection for high-level estrogen alleles began with the increased consumption of phytoestrogens in agricultural societies. The consumption of phytoestrogens in hunting-gathering societies, prior to agriculture, was low and did not create selective pressure for an increase in estrogen levels.

time in human evolution to increase the frequency of advantageous high-level estrogen alleles. Perhaps during evolutionary times lower levels of hormones were optimal? I suggest that they were indeed optimal for a long expanse of human evolution until lifestyles changed and new factors reducing the levels of hormones appeared.

It is likely that the most important lifestyle changes began with the introduction of agriculture. With agriculture, people began to consume, on a much larger scale, foods that contain high concentrations of chemicals that can interfere with steroid metabolism and physiological functions. Agriculture exposed women to higher concentrations of phytoestrogens in their diet than ever before.

Phytoestrogens, which are present in many kinds of agriculturally grown food, bind to estrogen receptors and, when consumed in high quantities, reduce the levels of endogenous estrogens. In addition, phytoestrogens exhibit far more potent binding abilities to estrogen receptors when the diet is also high in carbohydrates, which is a common feature of agricultural diets. Therefore, increased phytoestrogen consumption is likely to cause lower fecundity in women.

In other words, low-level estrogen alleles were the "ancestral condition" before the origin of agriculture, when our human ancestors lived as hunter-gatherers (Figure 2.2). In ancestral populations, the levels of hormones encoded by these low-level alleles were optimal for ovarian function and were related to a high probability of conception. When phytoestrogens came to be consumed in such high quantities that they reduced the levels of hormones, high-level genotypes became selectively advantageous.

Dietary Phytoestrogens

Many plants contain chemicals that can mimic the metabolic action of estrogens produced by the body (Knight and Eden 1995; Setchell and Lydeking-Olsen 2003; Dixon 2004) (Table 2.1). These chemicals, called phytoestrogens or plant estrogens, react with estrogen receptors and can also interfere with the steroid metabolic pathway. Phytoestrogens consumed by humans include flavonoids, lignans, and coumestans. Flavonoids are found in legumes, including soybeans, chickpeas, and lima beans. Lignans are present in flaxseed and sesame seed and in lower levels in almost all cereals (the most frequently found lignans are enterolactone and enterodiol). Coumestans are present in clover, alfalfa, and soybean sprouts. Many commonly consumed fruits and vegetables also contain phytoestrogens. Onions and apples have high quantities of quercetin. Cocoa and black and green teas have high concentration of catechins. St. John's wort *(Hypericium perforatum),* a herb widely used as a folk remedy for depression, contains several different isoflavonoids, including quercetin, amentoflavone, kaempferol, myricetin, quercitrin, isoquercetin, and rutin. (As a child, I observed my maternal grandmother making a tea from this herb, which she collected herself in rural southeastern Poland.)

Table 2.1.   Levels of isoflavones and lignans in food products.

Plant	Genistein	Daidzein	Secoisolar- iciresinol	Matairesinol
Soybean	993–3,115	413–2,205	<1–8	<1
Kidney bean	<1–19	<1–2	2–4	<1
Chickpea	3–8	<1–8	<1	0
Pea	<1	<1	<1	<1
Lentil	<1	<1	<1	<1
Kudzu root	467	7,283	<1	<1
Flaxseed	0	0	10,247	30
Sesame seed	<1	6	2	17
Wheat bran	<1	<1	3	0
Barley (whole grain)	<1	<1	2	0
Rye bran	0	0	4	5
Strawberry	0	0	33	<1
Zucchini	0	0	23	<1
Black tea	trace	trace	73	12
Green tea	trace	trace	75	5

Note: Values in nanomoles per gram dry weight (after Dixon 2004).

After ingestion, flavonoids are hydrolyzed in the intestines to the compounds daidzein and genistein. They may be absorbed or further metabolized into various chemicals, including equol (Setchell and Cassidy 1999). Equol has estrogenic potency that is an order of magnitude greater than the potency of daidzein, from which it is produced (Setchell and Cassidy 1999). The chemical structure of equol is very similar to that of estradiol.

To make things even more complicated, there is high individual variation in metabolism of flavonoids, and the rate of that metabolism depends on other components of the diet. A diet high in carbohydrates, which causes increased intestinal fermentation, is conducive to more intense metabolism of phytoestrogens. In such a bacteria-rich environment, production of equol is increased. Levels of urinary equol are very variable among individuals, depending on their diets. A person who eats a soy-rich diet may have approximately 100-fold higher levels of urinary equol compared with a person who eats few soy products.

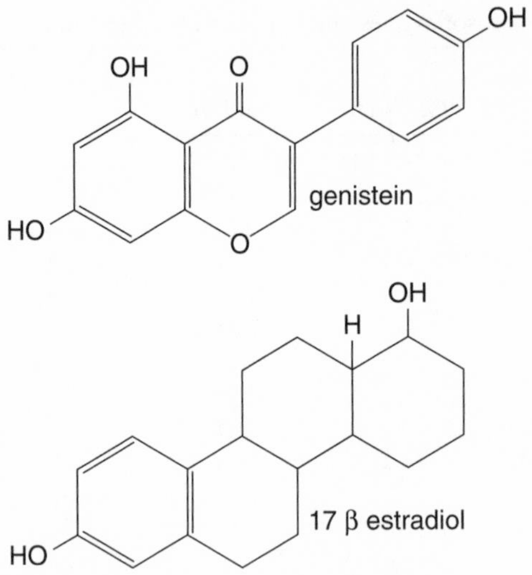

Figure 2.3. Chemical structures: phytoestrogen genistein versus estradiol produced by the ovaries.

Phytoestrogens: How Do They Work?

Most phytoestrogens have a chemical structure very similar to that of endogenous estrogens (Figure 2.3), so they can bind to the estrogen receptor. Phytoestrogens can also bind directly to the enzyme aromatase which converts androgens to estrogens, with a resulting decrease in estrogen production (Kao et al. 1998). Furthermore, phytoestrogens are thought to increase hepatic synthesis of sex hormone-binding globulin, which leads to a decrease in the amount of free and active endogenous estrogens present in the blood (Schmitt and Stopper 2001).

The success of phytoestrogens in producing estrogenic responses can be partially attributed to the uniqueness of the estrogen receptor among other steroid receptors. This receptor is referred to as "promiscuous," which means that it has the ability to interact with a wide variety of chemicals. It has been suggested that the large size of the binding cavity of the estrogen receptor, with a volume almost twice that of the estradiol molecule, is responsible for its "loose" binding behavior (Kuiper et al. 1997).

Phytoestrogens and "True" Estrogens

At the doses consumed by humans, phytoestrogens are much less potent than estradiol in inducing a biological response. Nevertheless, they seem to have the ability to lower circulating levels of estradiol. In other species, the effects of phytoestrogen-rich diets on reproduction are well-known. Reproductive suppression occurs in cattle and sheep grazing on pastures of phytoestrogen-rich clover and alfalfa. Captive cheetahs cannot produce young when they are fed a commercially prepared feline food with high levels of plant estrogens. In wild California quail, a diet rich in phytoestrogens, which are abundant in plants during dry seasons, suppresses reproduction (Leopold et al. 1976).

In humans, the effects of consumption of phytoestrogens are best known in postmenopausal women, in whom phytoestrogens appear to have a physiological estrogenic effect. It is important to note, however, that phytoestrogen consumption does not have exactly the same effect in postmenopausal and premenopausal women. In both groups, the phytoestrogens bind to estrogen receptors and induce the estrogenic effect on metabolism and physiology. However, in postmenopausal women, in whom only low levels of estrogens are supplied by the adipose tissue and ovarian production no longer occurs, consumption of phytoestrogens increases the overall levels of estrogens in the body (even though they are not "real" estrogens). In contrast, in premenopausal women, the "occupation" of estrogen receptors by phytoestrogens sends a signal that there are already some estrogens present in the body, which slows or switches off further production of estrogens by the ovaries. Because phytoestrogens and endogenous estrogens do not have the same potency for many aspects of physiology, including ovarian function, the levels of estrogens in the body may become too low when large amounts of phytoestrogens are ingested. This is especially true when endogenous estrogen production is controlled by low-level alleles, and thus these estrogens are already kept at relatively low levels.

In Southeast Asian women, who consume large amounts of soy products, estrogen levels in plasma are about 20 to 30 percent lower than those found in Western women. Japanese women also have, on average, longer menstrual cycles than women in Western countries (thirty-two days versus twenty-eight to twenty-nine days, respectively). Of course, factors other than diet could be also responsible for such differences. There are also observations, not confirmed by experimental studies, that

53

Southeast Asian women on a diet high in soy products are, to some extent, protected from breast cancer (Ganry 2002).

Does the consumption of phytoestrogen also have a physiological effect in premenopausal women? We would certainly expect to see lower estrogen levels in premenopausal women who have diets rich in these chemicals. One of the best dietary intervention studies, although with a sample size of only ten women, investigated changes in the concentrations of estradiol, progesterone, and gonadotropins (follicle-stimulating hormone and luteinizing hormone) and the length of the menstrual cycle in women on a soy-rich diet (Lu et al. 2000). Blood samples were collected daily from the ninth day of the menstrual cycle. Daily consumption of soy did not change the length of the menstrual cycle but did reduce the level of estradiol by 25 percent and level of progesterone by 45 percent. The substantial changes in steroid levels were not accompanied by changes in the level of gonadotropins, suggesting that the inhibitory effect of isoflavone consumption on steroid hormones is not mediated by gonadotropins. Perhaps soy isoflavones directly inhibit the steroid synthesis enzymes in ovaries. The decrease in the estradiol level was positively related to the level of isoflavones in urine but not to the level of consumed isoflavones, indicating that women clearly differ in their ability to systemically absorb consumed isoflavones. This suggests that individual variation in isoflavone metabolism might have played a role in this study.

In another dietary intervention study, estradiol levels in Japanese women decreased by 23 percent with increased soymilk consumption in comparison with their levels during menstrual cycles when soymilk was not consumed (Nagata et al. 1998). In that study, however, only a single blood sample was taken from two menstrual cycles. In a different randomized study, American women who had their diets supplemented with soy protein had reduced levels of estradiol, estrone, and estriol as measured in three urine samples collected during the midfollicular phase of their cycle (Xu et al. 1998). In comparison with a placebo group, premenopausal women consuming a dietary supplement of soy for a twelve-week period had decreased free estradiol and estrone levels, while their hormone-binding globulin level increased. In the women who were consuming the soy supplement, the mean menstrual cycle length increased by 3.52 days (Kumar et al. 2002).

These results suggest that intake of phytoestrogens reduces the level of endogenous steroids, but it should be acknowledged that not all studies confirm these effects. For example, a two-year dietary intervention study detected no statistically significant differences in the levels of steroids or sex hormone-binding globulin between a group of women consuming soy food compared with a group who ate no soy products (Maskarinec et al. 2004). A major problem with this and most other studies investigating the relationship between phytoestrogen consumption and steroids was insufficient sampling: steroids levels were measured in only one to two samples per menstrual cycle (Jasienska and Jasienski 2008).

Hormone levels were measured in a group of Polish women from whom saliva samples were collected for eighteen consecutive cycle days. The level of estradiol during their menstrual cycles was statistically significantly suppressed by means of tea consumption (Kapiszewska et al. 2006). Due to drinking black and green tea above the median level, the study group had a 25 percent reduction in estradiol levels. In addition, intake of epigallocatechin-3-gallate (EGCG), a catechin present mostly in tea, was also related to reduced estradiol levels. This relationship between tea drinking and estradiol levels is probably an example of phytoestrogens in action.

Although the catechins (EGCG and epicatechin gallate) present in green and black tea are not classic phytoestrogens, they too have an ability to bind to estrogen receptors. There are two types of estrogen receptors, and catechins show a greater ability to bind to beta-receptors than to alpha-receptors, behaving just like classic phytoestrogens. However, catechins in general have a much lower affinity for estrogen receptors than genistein and daidzein (i.e., soy phytoestrogens) (Goodin et al. 2002), so it is remarkable that their consumption can still affect estradiol levels in women.

Genetic Polymorphism and Levels of Reproductive Hormones

Coming back to the issue of polymorphism in genes involved in steroid hormone metabolic pathways, we need to ask the question of whether women with different *CYP17* genotypes differ in levels of steroid hormones circulating in their bodies. This question has only been recently addressed by researchers.

A cross-sectional study of 173 premenopausal women did not find any differences in estradiol levels among *CYP17* genotypes, but documented significant differences in the levels of the steroid hormone dehydroepiandrosterone (DHEA), a precursor for estradiol synthesis (Hong et al. 2004). Estradiol levels were measured in a single blood sample collected during the luteal phase between days twenty and twenty-four from the beginning of the cycle. Unfortunately, in reality, cycle phase cannot be reliably determined before the cycle has ended. As a consequence, it is likely that, for longer cycles, day twenty could still belong to the follicular phase or even represent a day before ovulation with the highest levels of estradiol for the cycle. This sampling methodology, common in medical and epidemiological studies, may lead to comparisons of women who are actually at different physiological points in their cycles.

Another study that did not find a difference in estradiol levels among *CYP17* genotypes also used a single blood sample for hormone measurements (Garcia-Closas et al. 2002). In addition, this study had relatively small sample sizes—for analysis of the early and late follicular levels of estradiol, it had only fifty-two and fifty-three women, respectively. Because the A2/A2 genotype was present in 16 percent of the women in this study group, the groups with A2/A2 genotypes used for estradiol comparisons probably had fewer than nine women. A large study of 636 premenopausal British women found no differences among *CYP17* genotypes in estradiol levels either (Travis et al. 2004).

Few other studies have shown variation in estradiol levels among the *CYP17* genotype. Estradiol levels measured around day eleven of the menstrual cycle were 11 percent and 57 percent higher among women with genotypes A1/A2 and A2/A2, respectively, compared with A1/A1 women (Feigelson et al. 1998). In the luteal phase, around day twenty-two of the cycle, the estradiol levels were 7 percent and 28 percent higher for women with at least one A2 allele. Among women with a lower body mass index (BMI of 25 kg/m^2 or less), those with the A2/A2 genotype had 42 percent higher estradiol levels than women with the A1/A1 genotype (Small et al. 2005). Heterozygotes had intermediate levels, 19 percent higher than A1/A1 genotype. However, among women with a higher BMI, the three *CYP17* genotypes did not differ in estradiol levels. It is possible that women with a higher BMI have higher levels of estradiol produced by adipose tissue, which may interfere with the production of estradiol by the ovaries. It should also be noted that both the studies described here obtained only one or two estradiol samples per woman. In fact, most studies have measured steroids in a single or only a few blood samples, even for premenopausal, cycling women; and, therefore, levels of estradiol may not be very reliable.

In a study on Polish women, we were able to sample estradiol across an entire menstrual cycle for each woman (Jasienska, Kapiszewska et al.

2006). We included samples from about eighteen consecutive cycle days (a few days from the beginning and a few from the end of each cycle were not included) for comparison of *CYP17* genotypes. In these women, the A2/A2 genotype had 54 percent higher estradiol levels than the A1/A1 genotype and 37 percent higher levels than the A1/A2 genotype. Out study is still in progress, and we had a relatively small sample size, but our estradiol measurements have been more precise than any of the other studies of *CYP17* and estradiol levels.

Polymorphism in Steroid-Producing Genes: Frequencies of Genotypes in Human Populations

We would like to explore the hypothesis that selection pressure resulting from changes in the patterns of food consumption that are characteristic of many agricultural societies have caused changes in the frequency of alleles of genes involved in steroid hormone production. First, we would need to document that there are differences in the frequencies of *CYP17* genotypes among populations. Second, we should show that these genetic differences are related to differences in consumption of phytoestrogens. In populations where the diet contains a lot of phytoestrogens, we would expect to find a higher frequency of high-level genotypes than in populations with lower phytoestrogen consumption. And this is exactly the pattern we observe.

First, the frequency of the high-level A2 allele varies among populations (Table 2.2, Figure 2.4). Only about 14.2 percent of Europeans (including people of European descent) and 13 percent of Africans (including people of African descent) have A2/A2 genotypes. The Hispanic population seems to have a slightly higher frequency of the A2 allele (more than 18 percent have the A2/A2 genotype), but there have been only two studies of this group to date. A much higher frequency is observed in Asian populations than in European or African groups. Almost 22 percent of the Japanese population has the A2/A2 genotype, and more than 33 percent of the Taiwanese population.

In light of the agriculture hypothesis discussed in this chapter, polymorphism of the loci involved in steroid metabolism within a population is not surprising. The origins of agriculturally grown food are relatively recent (no more than 14,000 years old), and natural selection may not have had enough time to replace low-level alleles even in populations with very high phytoestrogen consumption. Additionally, even within a

Table 2.2. Polymorphism in CYP17 gene in different ethnic populations.

Ethnicity	Study	Control/Cases	Number of individuals with CYP17 genotype			Frequencies of CYP17 genotypes (%)			Population details
			A1/A1	A1/A2	A2/A2	A1/A1	A1/A2	A2/A2	
African and of African descent	Feigelson et al. 1999		105	111	41	41	43	16	U.S.
	Kittles et al. 2001	Control	24	27	5	43	48	9	Nigeria
	Kittles et al. 2001	Control	55	46	10	50	41	9	African American
	Lai et al. 2001		36	31	11	46	40	14	Canada: nulliparous
	Lunn et al. 1999		51	46	18	44	40	16	U.S.
	Small et al. 2005	Control	40	49	14	39	48	14	Fragile X syndrome
	Weston et al. 1998	Control	15	18	2	43	51	6	U.S.: breast cancer
		Cases	7	10	3	35	50	15	
	Total		333	338	104	43.0	43.6	13.4	
European and of European descent	Allen, Forrest, and Key 2001		266	273	83	43	44	13	U.K.
	Ambrosone et al. 2003	Control	95	71	22	51	38	12	U.S.: breast cancer
	Bergman-Jungestrom et al. 1999	Cases	109	83	15	53	40	7	Sweden: breast cancer
		Control	53	55	9	45	47	8	
	Chang et al. 2001	Cases	32	62	15	29	57	14	
		Control	76	78	26	42	43	14	
	Cui et al. 2003		574	636	237	40	44	16	Australia

Diamanti-Kandarakis et al. 1999	Control	22	28	0	44	56	0	Greece: polycystic ovary syndrome
Dunning et al. 1998	Cases	17	29	4	34	58	8	U.K.: breast cancer
	Control	229	277	85	39	47	14	
Feigelson et al. 1998		28	45	10	34	54	12	U.S.
Feigelson et al. 1999		49	69	14	37	52	11	U.S.
Garcia-Closas et al. 2002		81	102	35	37	47	16	U.S.: premenopausal
Garner et al. 2002	Control	111	96	34	46	40	14	U.S.: ovarian cancer
	Cases	70	120	35	31	53	16	
Gudmundsdottir et al. 2003	Control	103	160	46	33	52	15	Iceland: breast cancer
Haiman et al. 1999	Cases	12	18	9	31	46	23	U.K.: breast cancer
	Control	217	307	94	35	50	15	
Haiman et al. 2001	Cases	178	212	73	38	46	16	U.K.: endometrial cancer
	Control	197	267	90	36	48	16	
Kittles et al. 2001	Cases	76	92	16	41	50	9	U.S.
	Control	28	38	8	38	51	11	
Kristensen et al. 1999	Control	74	101	26	37	50	13	Norway: breast cancer
Kuligina et al. 2000	Cases	202	241	67	40	47	13	Russia: breast cancer
	Control	61	77	44	34	42	24	
Lai et al. 2001	Cases	82	111	47	34	46	20	Canada: nulliparous
Lunn et al. 1999		126	171	48	37	50	14	U.S.
		47	48	20	41	42	17	

(continued)

Table 2.2. *(continued)*

Ethnicity	Study	Control/ Cases	Number of individuals with CYP17 genotype			Frequencies of CYP17 genotypes (%)			Population details
			A1/A1	A1/A2	A2/A2	A1/A1	A1/A2	A2/A2	
	McCann et al. 2002	Control	48	28	10	56	33	12	U.S.: breast cancer
		Cases	58	31	7	60	32	7	
	Mitrunen et al. 2000	Control	200	220	60	42	46	13	Finland: breast cancer
	Small et al. 2005	Cases	199	227	53	42	47	11	Fragile X syndrome
	Spurdle et al. 2000	Control	17	26	6	35	53	12	
		Control	115	139	44	39	47	15	Australia: ovarian cancer
	Stanford et al. 2002	Cases	118	150	51	37	47	16	U.S.: prostate cancer
		Control	188	256	79	36	49	15	
		Cases	228	248	84	41	44	15	
	Techatraisak, Conway, and Rumsby 1997	Control	61	54	9	49	44	7	Congenital adrenal hyperplasia or polycystic ovary syndrome
	Wadelius et al. 1999	Cases	67	63	2	51	48	2	
		Control	46	88	26	29	55	16	Prostate cancer
		Cases	70	74	34	39	42	19	
	Weston et al. 1998	Control	49	74	25	33	50	17	U.K.: breast cancer
		Cases	29	35	12	38	46	16	

	Young et al. 1999	Control	58	67	14	42	48	10	U.S.
		Cases	38	55	10	37	53	10	
	Zmuda et al. 2001		120	163	50	36	49	15	
	Total		4,924	5,865	1,788	39.2	46.6	14.2	
Hispanic	Feigelson et al. 1999		70	112	40	32	50	18	U.S.
	Weston et al. 1998	Control	28	19	10	49	33	18	U.S.: breast cancer
		Cases	9	12	6	33	44	22	
	Total		107	143	56	35.0	46.7	18.3	
Japanese	Feigelson et al. 1999		40	67	31	29	49	22	U.S.
	Gorai et al. 2003		101	118	31	40	47	12	Healthy women
	Habuchi et al. 2000	Control	33	62	36	25	47	27	Prostate cancer
		Cases	95	111	46	38	44	18	
	Hamajima et al. 2000	Control	44	95	27	27	57	16	Breast cancer
		Cases	41	83	20	28	58	14	
	Huang et al. 1999	Control	251	452	256	26	47	27	Rheumatoid arthritis
		Cases	113	173	90	30	46	24	
	Kado et al. 2002	Control	63	87	27	36	49	15	Endometriosis
		Cases	46	68	26	33	49	19	
	Miyoshi et al. 2000		48	106	41	25	54	21	
	Total		875	1,422	631	29.9	48.6	21.6	
Other East Asia	Chen et al. 2005	Control	9	60	33	9	59	32	Taiwan: oral cancer
		Cases	31	58	48	23	42	35	
	Huang et al. 1999	Control	28	63	35	22	50	28	Taiwan: breast cancer
		Cases	25	54	44	20	44	36	
	Lai et al. 2001		14	45	17	18	59	22	Canada: nulliparous

(continued)

Table 2.2. (continued)

Ethnicity	Study	Control/ Cases	Number of individuals with CYP17 genotype			Frequencies of CYP17 genotypes (%)			Population details
			A1/A1	A1/A2	A2/A2	A1/A1	A1/A2	A2/A2	
	Lo et al. 2005	Control	9	61	34	9	59	33	Taiwan: rheumatoid arthritis
		Cases	25	116	52	13	60	27	
	Lunn et al. 1999		26	54	30	24	49	27	Taiwan
	Wu et al. 2003	Control	109	333	229	16	50	34	Singapore, Chinese: breast cancer
		Cases	37	82	69	20	44	37	
	Yu et al. 2001	Control	37	111	90	16	47	38	Taiwan: hepatocellular carcinoma
		Cases	20	56	43	17	47	36	
	Total		370	1,093	724	16.9	50.0	33.1	

Note: Frequencies do not always add up to 100 percent because of rounding errors.

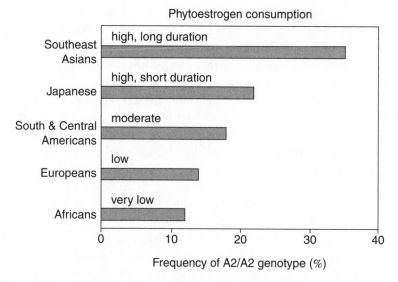

Phytoestrogen consumption

Figure 2.4. Population variation in the frequency of A2/A2 (high-estrogen-producing) *CYP17* alleles corresponds to the consumption of phytoestrogens (both in the amount and in historical duration).

population there is often considerable variation in dietary composition among social classes or geographic regions. Further, production of high levels of steroid hormones may not be free from metabolic costs, so it would not be promoted by natural selection unless ingested phytoestrogens have brought the levels of hormones down and consequently reduced fecundity.

In populations characterized by higher frequencies of high-level alleles, consumption of phytoestrogen-rich diets seems to have been introduced earlier in history, and phytoestrogen-rich foods are consumed in larger amounts than in populations with lower frequencies of these alleles. We will look at these relationships a little later in this chapter.

Evolution of the Human Diet: Phytoestrogen Consumption Increases

Our closest primate relatives almost exclusively eat plant foods. Although common chimpanzees occasionally eat insects and also can hunt and clearly enjoy eating meat, they only obtain about 5 percent of their food from meat sources. Intense hominid carnivory, either from hunting or scavenging, may have developed as early as 2.5 million years ago (de Heinzelin et al. 1999; Alvard 2001). In contemporary human populations, meat eating ranges from a few percent to almost 100 percent (in high-latitude hunter-gatherers such as the Inuit). Most contemporary populations of hunter-gatherers eat between 20 to 65 percent of their food as meat (Cordain et al. 2000), although some groups have lower meat consumption. In the !Kung of the Kalahari Desert in Central Africa, meat comprises 15 percent of all food consumed, and in the Efe pygmies of the Ituri Forest in the Democratic Republic of Congo only 9 percent of calories come from meat.

Development of agriculture about 14,000 years ago dramatically decreased dietary variability (dietary breadth). Most populations of Paleolithic humans probably consumed up to 100 different species of plants over the course of the year, but agriculturalists typically rely on a single species of cereal as the main dietary staple (Eaton, Eaton III, and Cordain 2002). The main staple has been rice in Asia, wheat in temperate Asia and Europe, millet or sorghum in Africa, and maize in the New World (Larsen 2002). In the majority of agricultural societies, dietary intake of meat is greatly reduced, and grains provide from 40 to 90 percent of caloric requirements (Eaton, Eaton III, and Cordain 2002).

The nomadic Turkana of East Africa, who are animal herders, have an unusually high consumption of animal products. During the periods when milk is abundant, people eat very little other food. A person who has a diet of 1,600 kilocalories per day may eat as much as 1,400 kilocalories per day in the form of milk. In seasons when animals do not produce much milk, milk consumption provides about 700 kilocalories per day and the rest of a person's calories come from meat, blood, cereals, and some wild foods (Little 2002; Lips 2007). Even in pastoralist populations, consumption of nonanimal products can be high. Among the Phala nomads of Tibet, 50 percent of dietary calories comes from grains (Goldstein and Bell 2002), and the Evenki herders of Siberia derive as much as 70 percent of dietary calories from nonanimal foods (Leonard et al. 2002).

Domesticated legumes contain high levels of plant estrogens; peas, chickpeas, lentils, and soybeans are especially relevant here because of their large concentrations of genistein and daidzein (Dixon 2004). These legumes were known to people from the very beginning of food production. In the Near East, in the Fertile Crescent—the cradle of agriculture—archeological evidence suggests that ancient farmers were growing legumes as early as 13,000 years ago (Abbo et al. 2003). Flax, an important source of lignans, was domesticated around 7000 BC; flax fiber has been used in the production of textiles, and flax seeds are a rich source of oil consumed in many parts of Europe (Diamond 1997, 125–126). Of course, even before agriculture people consumed plants that contained phytoestrogens, but it is unlikely that such plant species were a significant part of their overall diet given their wide dietary breadth, seasonally changing plant consumption, and lack of plant food storage.

Many plants common in the diets of agricultural and industrial societies contain some phytoestrogens, but their content is highly variable (see Table 2.1). All cereals contain lignans but in relatively low quantities in comparison with flaxseed. Similarly, all legumes contain isoflavones, but only soybean has very high levels of these phytoestrogens. Traditional Southeast Asian diets, especially in Japan, China, and Indonesia, are characterized by a much higher consumption of phytoestrogens than the European diet, mostly owing to the use of soybeans as the main source of dietary protein. Neither soybeans nor flaxseed are important components of contemporary European diets. However, flaxseed as the main source of dietary oil was probably important in early agricultural diets in Europe, but still, consumption of phytoestrogens was much lower than in Asian diets.

In early agricultural diets in Africa, people most likely did not consume high quantities of foods with a high phytoestrogen content. In the New World, maize was the main dietary staple, and there was also a relatively high consumption of legumes. Based on this very rough estimation of the phytoestrogen content in diets of historical agricultural populations, some predictions about genetic polymorphism in *CYP17* can be proposed.

The Phytoestrogen–*CYP17* Link

High-level estrogen alleles (A2) should be least frequent in Africa, followed by Europe, the New World, and Asian populations, with the highest levels in Japan, Taiwan, and Singapore (see Table 2.2, Figure 2.4). Phytoestrogen consumption in Japan and Taiwan was most likely quite similar; however, the origin of agriculture was more recent in Japan than in Taiwan. In Taiwan, agriculture already existed before 3000 BC. Agriculture was brought to Japan by Korean farmers around 400 BC. The modern Japanese population, as suggested by skeletal and DNA evidence, is a hybrid of these Korean migrant farmers and the early hunter-gathering inhabitants of the Japanese islands (Diamond and Bellwood 2003).

Korean farmers had some exposure to a phytoestrogen-rich diet and could have introduced high-level estrogen alleles to the population of Japanese hunter-gatherers. A diet rich in soybeans exerted further selective pressure on Japanese women.

Contemporary Southeast Asian diets have a much higher concentration of phytoestrogens than European diets, but there is variation among Southeast Asian countries. Consumption of isoflavone from soy products is estimated to be 15 to 45 milligrams per day for Japanese and Chinese adults (Setchell 2001). Indonesians have an even higher consumption of soy; their consumption of isoflavone is on the order of 150 milligrams per day, but data on *CYP17* polymorphisms are not available for the Indonesian population. By contrast, isoflavone intake in the British diet is less than 1 milligram per day. In Southeast Asian women, the circulating plasma levels of phytoestrogens are approximately thirty times higher than in women who consume a typical Western diet (Setchell and Lydeking-Olsen 2003).

An important issue to address here is whether there has been enough time since the beginning of agriculture for the rise of polymorphism in genes related to steroid synthesis. The best example of selection for traits that became important after the invention of agriculture is the ability to metabolize milk by human adults (Holden and Mace 1997). Human infants use the enzyme lactase to break down the milk sugar lactose; after weaning, lactase levels drop in many ethnic groups (when genes responsible for lactase production become switched off), and people can no longer digest milk effectively (Fuller 2000). Hunter-gatherers did not need to digest milk after weaning because without domesticated animals they had no sources of milk.

It has been suggested that the ability to retain lactase synthesis after weaning appeared in humans when the domestication of animals added a good source of calcium to the diet. The ability to consume dairy products clearly increased the fitness of people who had the trait (Bamshad and Motulsky 2008). It was especially important for populations in northern latitudes where a short growing season did not allow sufficient calcium to be obtained from leafy green vegetables. Not surprisingly, lactase persistence, a dominantly inherited genetic trait, occurs with high frequency in northern Europeans (Swallow 2003).

One may ask why other hormone-suppressing environmental factors did not have a selective effect on genes. Many factors are capable of suppressing ovarian function, so why is only the suppressing effect of con-

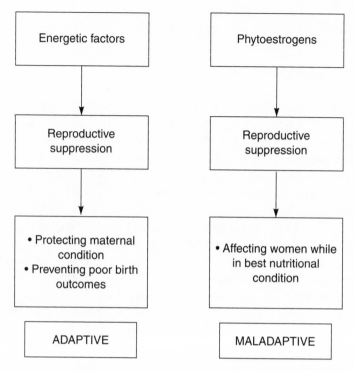

Figure 2.5. Reproductive suppression in women in response to low energy availability is adaptive, but reproductive suppression due to high consumption of phytoestrogens is maladaptive.

sumption of phytoestrogens postulated to change allele frequency? Factors related to energy availability (poor energy diet, intense physical activity, weight loss) suppress ovarian function, but this suppression, as discussed in Chapter 1, is adaptive. It prevents a woman from devoting time, energy, and nutrients to reproduction when her nutritional condition is poor (Figure 2.5). In contrast, reproductive suppression caused by dietary phytoestrogens is not adaptive—it does not confer a selective advantage. Phytoestrogens most effectively suppressed reproductive function during times of high food availability. The highest quantities of phytoestrogens were probably consumed after harvests when food in general was abundant. Therefore, phytoestrogens caused reproductive suppression in women during the best time for them to become pregnant. Such suppression was clearly maladaptive—it makes no sense, evolutionarily speaking.

Phytoestrogens, Testosterone Levels, and
Allele Frequencies in Human Males

Even in populations with a long history of high phytoestrogen consumption, frequencies of high-level alleles are relatively low. What are the causes of the low prevalence of such alleles? It is possible that the selective pressure imposed by phytoestrogens on *CYP17* genes was not very strong, so not enough time (measured in human generations) has elapsed for more change in the genetic composition of the population to occur. However, it is also likely that human males are to be "blamed" for the low frequency of these high estrogen-producing alleles.

When diet changed with the introduction of agriculture, it changed also for males. Phytoestrogens also have an ability to affect levels of male steroid hormones, including testosterone. Enzyme P450c17a, encoded by the *CYP17* gene, is also involved in steroid synthesis in males because it mediates the activities of two enzymes (17-hydroxylase and 17,20-lyase) at key points during the testosterone biosynthesis in the testes (Ntais, Polycarpou, and Ioannidis 2003). The same mutation in *CYP17* (a replacement of nucleotide thymine with cytosine: allele A2) that is associated with high levels of estrogen in females is associated with high levels of testosterone in males. It has been suggested that this allele increases levels of testosterone via an increased rate of gene transcription. In vitro studies suggest that physiological concentrations of the phytoestrogen genistein cause down-regulation of testosterone and other androgen receptors. This is similar to effects described earlier for human females.

There are, however, no studies in vivo to show that phytoestrogens can indeed affect testosterone concentrations at the receptor level. In a dietary intervention study, soy supplementation resulted in reduced testosterone levels in men (unpublished data from a study conducted by VanVeldhuizen et al. cited in Holzbeierlein, McIntosh, and Thrasher 2005). The number of studies on *CYP17,* androgen levels, and phytoestrogens in men have been rapidly growing due to the potential role of plant estrogens in the prevention of prostate cancer (Hussain et al. 2003; Kumar et al. 2004; Lund et al. 2004; Holzbeierlein, McIntosh, and Thrasher 2005).

Because phytoestrogens may have a suppressive effect on testosterone levels, can we postulate that there was a similar selective pressure on the genetic polymorphism in the *CYP17* gene in human males as has been hypothesized for human females? Was an increase in phytoestrogen consumption in males an additional selective force for increasing the frequency of high-level *CYP17* alleles in human populations?

Testosterone in males is involved in reproductive physiology and is important for sperm production and sexual libido. However, testosterone also has important metabolic functions. Significant differences in testosterone levels are found in men from different populations (Ellison et al. 2002). In populations where nutrition is poor, men have lower testosterone levels than men from populations where food is abundant. Richard Bribiescas (2001, 2006) has suggested that because testosterone promotes synthesis of muscle tissue, men from populations with poor nutrition have reduced testosterone levels as a way of reducing expensive muscle synthesis. Muscle has high metabolic costs to maintain, so people with poor nutritional status are better off with fewer expensive muscles. Energy and protein instead may be used for other physiological functions such as maintaining the immune system.

Agriculture caused not only a reduction in dietary breadth with an increased consumption of phytoestrogens but—and this may come as a surprise to many—was associated with a deterioration in nutritional status. Early agriculturalists had a diet insufficient in both energy and specific nutrients, and, as a result, they had much poorer health than hunter-gatherers (Angel 1984; Cohen and Armelagos 1984; Larsen 1984, 1995; Fleming 1994). For men, early agriculture clearly was not the time for maintaining large amounts of expensive muscular tissue. In addition, during the same period, increased population density, housing, and exposure to new parasites from domesticated animals led to increased costs of maintaining the immune system. Men not only had less energy from diet but also increased energetic expenses.

Consumption of phytoestrogens, which reduce the levels of estrogen in women and testosterone in men may have different, gender-specific effects on the evolution of *CYP* genes polymorphism (Figure 2.6). In women, a reduction in estrogen levels has detrimental effects on fecundity, and thus a deteriorating effect on fitness. By contrast, in men, reduction in testosterone levels allows more energy for uses other than muscle synthesis and maintenance, such as promoting immune function and thereby increasing fitness. Consequently, it is possible that selection promoted high-level *CYP17* alleles in women but low-level alleles of the same gene in men. Evolutionary biologists have described many such traits that are selected in opposite directions in the two sexes—a phenomenon called the *intralocus*

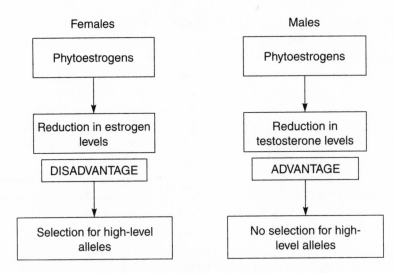

Figures

| Females | Males |

Females

Phytoestrogens

Reduction in estrogen levels

DISADVANTAGE

Selection for high-level alleles

Males

Phytoestrogens

Reduction in testosterone levels

ADVANTAGE

No selection for high-level alleles

Net outcome: a relatively weak selection pressure

Figure 2.6. A schematic model showing that high phytoestrogen consumption increased selective pressure for an increased frequency of alleles encoding high estrogen levels in human females, but no such pressure was present in human males.

sexual conflict (Rice 1998; Chippindale, Gibson, and Rice 2001; Morrow, Stewart, and Rice 2008).

The change in human diet that occurred with the advent of agriculture might have influenced the frequencies of alleles involved in steroid production, and this may help to explain why populations differ in the frequencies of high-level estrogen alleles. Genetic variation, together with variation caused by lifestyle factors (as discussed in Chapter 1), contributes to the variation in levels of reproductive hormones we observe today among women and among populations.

It is only fair to say that some evidence presented in this chapter still requires further research to be confirmed. Some studies do show that consumption of phytoestrogens reduces endogenous estrogen levels, but other studies have not found similar effects. Even less is known about the interaction of phytoestrogen consumption and testosterone levels in males. A more rigorous test of the hypothesis presented in this chapter would require that the relationship between the history of phytoestrogen consumption and the polymorphism of genes related to estrogen meta-

70

bolism be supported in a greater number of populations. Finally, polymorphism of other genes, not only of *CYP17*, should also be studied in this context.

We have discussed the most important factors that cause variation in the levels of reproductive hormones in women. The hypothesis of adaptive ovarian suppression can explain why women's physiology suppresses hormone levels in response to unfavorable conditions. The hypothesis of increased consumption of plant estrogens suggests a possible explanation for why genetically determined variation in hormone levels exists. In recent years, it has become clear that to fully understand variation in hormone levels we need to look not only at our genes and adult environment but also at environmental conditions experienced during fetal and childhood development.

In Chapter 3, we will ask whether ovarian function, just like many other physiological functions, is affected by the quality of the fetal environment. We will try to understand why the functioning of reproductive physiology in adulthood may be influenced by conditions experienced so much earlier in life. It is possible, as has been suggested by some researchers, that during its development the fetus receives information about the state of the environment filtered, of course, through the maternal physiology. Based on this information, the metabolism and physiology of the fetus are "programmed" in preparation for "predicted" future conditions. Such programmed physiology will provide a good fit for postnatal environmental conditions—assuming, of course, that the predictions made during the fetal period are correct.

3

YOU ARE WHAT YOU EAT . . . AS A FETUS

A connection between fetal development and reproductive physiology in adulthood is not surprising. Poor nutrition while in utero has a profound influence on the subsequent condition of the individual—a phenomenon sometimes called "fetal programming." Especially well proven is the relationship between undernutrition experienced early in life and an elevated risk of several metabolic diseases such as diabetes, cardiovascular disorders, and stroke (Barker 1994). Some physiological mechanisms leading to such adverse outcomes are already understood, mostly thanks to animal studies. Poor nutrition in utero causes changes in the physiology and metabolism of the organism that seem to be permanent. In response to poor nutrition, the size and structure of the internal organs may change, and thus many metabolic processes are affected. One of the best known examples is the case of insulin metabolism.

Individuals born after an undernourished pregnancy often in later life have muscles that are less sensitive to the hormone insulin. In healthy people, insulin causes the cells of many organs, including the muscles, to take up glucose from the blood. Glucose, a simple sugar that is a final product of food digestion, is a source of energy. When muscles become insulin resistant, their cells are reluctant to take up glucose, and too much glucose is left in the blood. High circulating blood levels are, in the long run, detrimental to the overall physiology of the body. Insulin resistance has been linked to a whole cluster of diseases (together called the metabolic syndrome). These include obesity, adult-onset diabetes (also called type 2 diabetes), hypertension, and cardiovascular disorders.

For a long time, these diseases were thought to originate from our lifestyle. Now we know that adult lifestyle can be only partially blamed for the metabolic syndrome and that early developmental conditions are also important. Since the 1980s, a rich set of data has documented the impact of uterine conditions on many aspects of animal metabolism, physiology, and health, and comparable results have been replicated by studies in several human populations (Barker 1994; Gluckman and Hanson 2005).

Although the importance of this area of research for human health is unquestionable, more detailed discussion is beyond the scope of this book. We limit ourselves to just *two* aspects of the long-term consequences of uterine environment. First, we will examine the evidence suggesting that the biology and health of an individual are affected by the conditions experienced by his or her ancestors. By this, I do not mean distant ancestors from the human evolutionary past but more recent generations of ancestors: mothers, grandmothers, and great-grandmothers.

Then we will look at the relationship between uterine conditions and reproductive function in women. Some studies have suggested that reproductive physiology may undergo permanent changes due to poor fetal development. I will not argue that these changes are pathologies— quite the opposite. I will ask whether these irreversible modifications in reproductive physiology that occur so early during development can be adaptive. In other words, is it possible that such early occurring reproductive "programming" may have a positive impact on adult reproductive physiology? Can this impact be so great that it actually increases the reproductive success of the individual?

The Dutch Famine

In humans, powerful, conclusive studies on the long-term effect of poor nutrition during gestation are difficult to undertake, partly because such experimental studies would not be ethical. There are many undernourished populations where women do not have adequate nutrition during pregnancy, but birth weight, the main indicator of nutrition in utero, is also affected by maternal condition *before* conception, including the mother's nutrition, workload, and disease burdens. In undernourished populations, all these factors confound the effect of poor nutritional status of the mother *during* pregnancy.

Further, in these populations, the impact of poor uterine nutrition on adult condition (which is the main interest of epidemiological studies) is also masked by the confounding influence of many other factors. For many people, poor uterine nutrition is just the beginning of lifelong nutritional deprivation. Poor nutritional conditions in childhood and adult life as well as a high incidence of infectious diseases all affect the health of an adult, and disentangling the effects of early and subsequent environments is very difficult or even impossible. It is not surprising, therefore, that well-nourished populations who went through "natural" experiments of short-term undernutrition, usually imposed by war, are of great interest.

The Dutch famine that occurred in the German-occupied Netherlands during the winter of 1944–1945 provided, many years later, an opportunity to study the long-term outcomes of short-term nutritional deprivation during fetal life. Problems with food began when the Germans banned all food transportation (Roseboom, de Rooij, and Painter 2006). This embargo was lifted in early November 1944, and food transport by canals and waterways was again permitted, but that did not help much because winter that year came early and was uncommonly severe, causing all waterways to freeze and making transport of food impossible. As a result, official daily rations for adults fell below 1,000 kilocalories in late November 1944. Between December 1944 and April 1945, adults received only between 400 and 800 kilocalories per day, which is between 20 and 40 percent of today's recommended level for a healthy adult woman. In the beginning, pregnant and nursing women received additional food, but such supplementation became impossible as the famine progressed.

During the winter, famine was not the only problem: there was no fuel, no gas or electricity, and in some areas no running water. Remarkably, the doctors and midwives providing medical care to pregnant women kept detailed written records about the pregnancies and deliveries, and about the size and health of the babies. These data are particularly well suited for an investigation of the long-term effects of in utero conditions on the health of adults because other aspects of the Dutch situation were equally unique.

The population who went through these dramatic events was very well nourished before the winter of 1944–1945. Similarly, the nutritional condition of Dutch people improved immediately after the liberation of the Netherlands, when official rations increased to about 2,000 kilocalories and remained sufficient thereafter. During the famine, the official rations for children under one year old never fell below 1,000 kilocalories; thus, after birth the children were not nutritionally deprived, at least in terms of their energy requirements (Roseboom, de Rooij, and Painter 2006).

In addition, direct comparisons are possible from the same population for people who were born just before the famine versus those conceived after the famine, as well as comparisons between groups of people who were exposed to the famine at different stages of their fetal development.

People exposed to famine during any stage while in utero had an increased risk of developing problems with insulin metabolism when they were in their fifties. Exposure to famine during early gestation was associated with a higher risk of coronary heart disease, disturbed blood coagulation, and obesity as well as an increased responsiveness to stress. Exposure at midgestation subsequently caused obstructive airways disease (Roseboom, de Rooij, and Painter 2006).

Birth Weight and Its Determinants

The best marker of the quality of the fetal environment is the infant's size at birth, which is why most studies use birth weight as an indicator. Low birth weight often indicates that physiological and metabolic adjustments occurred during fetal life. Birth weight is a consequence of not just of the maternal energy intake during pregnancy: adequate maternal intake, as observed in women from developed countries, is necessary for a normal birth weight, but supplementation of undernourished pregnant women in nonindustrialized countries has had a surprisingly limited impact on the birth weight of their children. Classic studies conducted in the rural Gambia (which will be described in more detail in Chapter 11) revealed that substantial supplementation during pregnancy results only in a small increase in birth weight (Prentice et al. 1987). Similarly, in Guatemala, supplementation of 10,000 kilocalories to pregnant woman increased the birth weight of their children by just 29 grams (Lechtig et al. 1978). A review of thirteen supplementation studies shows that balanced protein/energy supplementation of women during pregnancy has resulted in very small, biologically insignificant increases in the birth weight (25.4 grams on average) of their children (Kramer 1996).

The modest effect of supplementation during pregnancy on birth weight should not be interpreted as evidence that maternal nutrition is not important. Rather these results suggest that, in women who are generally undernourished, additional energy from supplements is not used for improving the biological quality of their current offspring. Women who are undernourished during pregnancy, who often live in less developed countries, are just as likely to have been undernourished before their pregnancies—and during most of their childhoods and adult lives. Some were probably born as small babies themselves as a result of having had undernourished mothers.

The child's birth weight shows little improvement when additional energy is received by the mother during pregnancy, but birth weight is related to maternal weight gain during pregnancy and her prepregnancy nutritional status. Remarkably, there is even a relationship between the birth weight of the mother herself and that of her baby (Bakketeig, Hoffman, and Harley 1979; Magnus, Bakketeig, and Skjaerven 1993; Selling et al. 2006). In fourteen studies on this issue, for every 100 grams of increase in maternal birth weight, the birth weight of a woman's child increased by 10 to 20 grams (Ramakrishnan, Martorell et al. 1999).

In part, this relationship can be explained by the presence of common genes shared by the mother and her offspring. We know that genes are important because paternal birth weight also positively correlates with the birth weight of the child (Ramakrishnan, Martorell et al. 1999). However, the magnitude of the relationship is greater for maternal birth weight than paternal birth weight, suggesting that the effect of the maternal environment is important as well. Recent studies on human pregnancies originating from egg donation show that the child's birth size is more strongly correlated with the body height of the egg recipient than with the height of the egg donor (Brooks et al. 1995), again emphasizing the importance of the maternal environment.

Intergenerational effects on size at birth are not limited to a correlation between the birth weight of the mother and her offspring. The birth weight of the child depends also on its mother's growth during her own childhood. Mothers who are taller as children subsequently give birth to heavier babies, and this relationship does not change after controlling for the mother's own birth weight and her adult height (Martin et al. 2004). In fact, a child's birth weight is best predicted by the mother's leg length, which is a sensitive indicator of her nutrition during childhood (Gunnell 2002).

Intergenerational effects can span several generations (Ounsted 1986; Ounsted, Scott, and Ounsted 1986; Emanuel, Kimpo, and Moceri 2004) A baby's birth weight was partially determined before the baby's own mother was even conceived! A mother's birth weight is determined by her mother's birth weight as well as her mother's childhood growth and adult stature (Emanuel, Kimpo, and Moceri 2004). Therefore, a baby's birth weight is in part determined by the grandmother's birth weight, but also very likely by the great-grandmother's birth weight (since the grandmother's birth weight depended on the great-grandmother's birth weight). We do not know for how many generations this effect really persists because it is hard to obtain reliable data on birth weights for multiple generations of families.

In summary, intergenerational nutritional effects are clearly important determinants of birth weight of a child, perhaps even more important than the maternal nutrition it receives during pregnancy.

Adaptations or Developmental Constraints?

The consequences of developing in an unfavorable uterine environment are relatively well known (Barker 1994; Gluckman and Hanson 2005), but there is outgoing debate as to *why* such changes in physiology and metabolism occur at all. A simple explanation would be that physiological changes occurring during fetal life result from developmental constraints. Perhaps in an environment deprived of sufficient energy and nutrients small body size and changed physiology are just the best that can be achieved given the circumstances (Bateson 2001). In other words, there is nothing positive about being small and having an altered physiology— these are just unavoidable consequences of developmental hardship.

Alternatively, physiological and metabolic changes may be functional, preparing the individual for a future life in an equally energy-poor environment. For this reason, the medical and epidemiological literature often refers to such changes in fetal physiology and metabolism as "adaptive," following the work of Hales, Barker, and others (Hales and Barker 1992; Barker 1994). It is important to realize, however, that this line of reasoning is *highly* speculative: we do not know whether permanent changes in, for example, insulin metabolism or renal function are really adaptations. That is, we do not know whether they enhance the survival or reproductive success of the affected individual.

Physiological changes occurring in poor fetal conditions seem to lead to the development of an individual who has a so-called *thrifty phenotype*. The thrifty phenotype hypothesis was put forward to explain the observed relationship between poor fetal growth and the increased risk of problems with glucose metabolism (Hales and Barker 1992, 2001). The hypothesis proposes that an individual developing in poor nutritional conditions is born better equipped to survive in later life on similarly limited resources. For example, having muscle cells that take up less glucose, which is available in limited amounts for poorly fed individuals, saves some glucose for other physiological needs.

According to this hypothesis, a physiology altered (or programmed) in utero this way is advantageous or "adaptive" only if the poor nutritional conditions are experienced consistently during an entire life. When subsequent, postnatal nutritional conditions significantly improve in comparison with the conditions experienced during fetal development, metabolic diseases such as diabetes often develop. However, there is no convincing evidence to prove that having this programmed physiology even in poor postnatal environments is truly adaptive—that is, that it leads to higher evolutionary fitness.

It is important to note that, with the exception of participants in experimental studies, fetal developmental conditions are usually unknown. The majority of data about the impact of human fetal conditions comes from retrospective studies. The information we have about nutrition during pregnancy has been inferred from the birth size of babies. Small newborn size does indeed often indicate that the mother underwent nutritional stress during pregnancy, but other factors such as smoking or infection may also restrict growth in utero.

In this chapter, I will discuss the adaptive significance of fetal programming from an evolutionary perspective, so I will be mostly interested in small birth size as an indicator of poor nutrition during pregnancy. Factors to which humans had little exposure during their pre-agricultural history (such as smoking) could not have had any impact as an evolutionarily relevant selective pressure.

It is also important to keep in mind that small birth size per se is not a cause of health problems in adulthood. It is, in fact, difficult to determine whether small birth size by itself is an interesting variable, or whether it has fitness costs or potential benefits for an adult. Small size at birth is probably merely an indicator of nutritional (or other) problems experienced during fetal life and, as we have reviewed recently, during several recent ancestral generations. It may be a marker of internal anatomical, physiological, and metabolic adjustments that occurred in utero. In other words, small size and these adjustments are all consequences of in utero nutritional deprivation. Would a hypothetical person who was small at birth but with none of these adjustments be any different in terms of the risk of adult metabolic diseases from a person born with a larger size? Most likely not.

We say that small size at birth is a risk factor for diabetes in adult life, but we know that the direct reasons for increased risk are changes in insulin metabolism. Such changes do not occur as a consequence of small birth size, but they are determined by the same factors that reduce body growth in utero and thus cause the small size. Similarly, women who were small at birth have a reduced risk of breast cancer. Again, small size at birth per se is unlikely to contribute to this relationship. Women born small apparently have adjustments to their reproductive physiology, as will be discussed later in this chapter, such that they produce lower levels of estrogens in menstrual cycles. Lower exposure of the breast to estrogens during one's lifetime reduces the risk of cancer.

The Intergenerational Signal of Environmental Quality

The significance of fetal programming has been an intensely debated subject, specifically whether the physiological and metabolic modifications observed in people of small birth size are indeed adaptive (Barker 1995; Wells 2003; Kuzawa 2004; Gluckman and Hanson 2005). The notion of adaptive significance has been discussed mainly in relation to adult body size, nonreproductive physiology, and senescence. Small body size and a modified metabolism are supposed to be adaptations to anticipated nutritional stress later in life (Bateson 2001; Hales and Barker 2001; Bateson et al. 2004; Gluckman and Hanson 2005). To evaluate the hypothesis that the uterine environment prepares individuals for lifelong environmental conditions, we need to assume that conditions during pregnancy are reliably correlated with future conditions, that is, that they have some predictive power of "telling" the fetus that its postpartum life will be generally poor or rich in nutrition. However, pregnancy in humans lasts for a relatively short time in comparison to the long human life span. If environmental conditions are stable, it is more likely that conditions experienced by the mother during pregnancy will be similar to those present for the next several decades.

We do not know for sure how stable the environment was during human evolutionary history, but there are reasons to believe that there were substantial fluctuations of such duration that different environmental conditions were experienced during a single human lifetime. Paleoclimatic evidence suggests that local climate and ecology have been relatively stable for the last 10,000 years, but not before that. During the past 100,000 years the climate was characterized by substantial, short-term variability (Burroughs 2005, 18–73).

The concept of an intergenerational signal (Kuzawa 2005) attempts to deal with this problem by extending the fetal nutritional experience from just nutrition during pregnancy to the integrated experience of environmental quality during several past generations. We have already looked at some evidence that birth size and related physiological and metabolic adjustments result not only from maternal nutrition during pregnancy but also from maternal and grandmaternal birth weights and nutritional conditions. Such integrated signals are capable of providing a much more reliable assessment of the quality of environmental conditions than signals based only on the condition of the mother during pregnancy.

The physiological and metabolic mechanisms responsible for intergenerational effects are not well understood, but the rapidly developing field of epigenetic inheritance has suggested that intergenerational information about environmental quality can be passed to the following generations by changing gene expression (Jablonka and Lamb 2005). Therefore, information about environmental quality such as nutrition, workload, and diseases can be passed along generations (Kuzawa 2008).

But to complicate matters even further, we need to remember that during pregnancy all information the fetus receives comes from the mother anyway. We assume that the mother has the "knowledge" of the integrated assessment of past nutritional conditions that she, in turn, received from her mother, and that she is physiologically able to pass this signal to the fetus. But, evolutionarily speaking, is sharing this information in her best interest?

The Conflicting Interests of the Mother and Her Fetus

Most proposed scenarios about the adaptive significance of fetal programming are problematic because they only talk about the evolutionary fitness of the fetus and ignore the fitness of the mother. Developmental effects should be discussed in the context of mother-offspring conflict, as has been elegantly done by Jonathan C. K. Wells (2003), based on the theoretical framework developed by Robert Trivers (1974) and extended by David Haig (1992). The interests of the mother and the fetus are to a large extent exactly the same. Both parties are interested in the fetus developing in a way that will ensure its survival and high future reproductive success. The mother's fitness rises when the fitness of her offspring is high—the higher the reproductive success of her offspring the more of her genes will be present in next generations. However, the mother's fitness integrates all her reproductive episodes, so it is best served by a long-term reproductive strategy.

In a long-lived species such as humans, a female has about thirty years during which she can bear children. Emily Lennox, an Irish aristocrat living in the eighteenth and nineteenth centuries, had twenty-two children during her reproductive span (Tillyard 1994). Emily could afford the energetic costs of this intense reproductive effort because she was in good nutritional condition and, like many aristocratic women of her time, did not breastfeed her own children but rather relied on wet nurses. During human evolutionary history, most women did not enjoy such luxuries. When resources are limited, life-history theory predicts trade-offs between the current and future reproduction. The woman needs to allocate energy and nutrients to the fetus and later to the nursing infant, but also to herself. If she invests too much in the current offspring, she may be at risk of so-called maternal depletion, which may negatively impact her future reproductive prospects.

The offspring does not appreciate such restricted maternal allocation of nutrients and energy. Even though the offspring will be genetically related to its future siblings (and having siblings is important for its inclusive fitness), the offspring has both maternal and paternal genes. Maternal genes certainly will be shared by all siblings, but that is not necessarily the case with paternal genes. The future children of the mother may be fathered by different men. Therefore, maternal and paternal parts of the genome in a given offspring have different and conflicting interests (Haig 2008).

The maternal half of the fetal genome may be more willing to follow the maternal strategy: fewer maternal resources allocated to the fetus allows the mother to make more children, who will have genes in common with the fetus. The paternal half of the genome is far more selfish. It is in the interest of paternal genes to get as much of maternal resources as possible, even though this strategy may lead to deterioration of maternal conditions and reduce her future reproductive potential.

Is Fetal "Programming" Benefiting the Mother?

The maternal fitness model proposed by Jonathan C. K. Wells (2003) suggests that fetal programming is more beneficial for the mother than for the offspring. In long-lived species such as humans who have long-dependent

81

offspring, mothers not only invest energy and nutrients during gestation and lactation, but also must also provide nutrition for many years after weaning. Mothers with poor nutritional status, and especially those with a long intergenerational history of poor nutrition among their ancestors, may benefit from having offspring with lower nutritional demands. Therefore, says Wells, "programming of the offspring may be seen as the outcome of competition between maternal and offspring strategies, with offspring size, body composition and metabolism all manipulated by the mother according to the quality of the environment." This hypothesis, Wells argues, explains why permanent changes in the organism occur so early during the development, while the other "thrifty phenotype" hypotheses fail to explain such early timing.

Theoretically, an ability to maintain plasticity in physiology after birth would be more beneficial for the individual than making irreversible changes to metabolism at such an early stage of life. It is likely that this timing results from time constraints on the developmental schedule—that is, that decisions about size and the function of specific organs can only be made at a particular time. Wells argues, however, that such early timing of permanent adjustments is beneficial for the mother, giving her a guarantee that she will be able to meet the future nutritional requirements of her child. She may control how much energy and nutrients are passed to a given offspring during pregnancy and lactation; after weaning, she loses such physiological control. By programming a small-size child with a thrifty metabolism, she can exert her control over its nutritional needs also for many years after weaning, when the child still depends on her for supplies of energy and nutrients. In poor environmental conditions, a mother prefers a thrifty, and thus less energetically expensive, child.

Is it in the child's interest to have a thrifty metabolism? The hypotheses on fetal programming assume that having a thrifty phenotype is beneficial in poor environmental conditions, but we do not have convincing evidence that such a strategy really helps individuals survive better at a young age or that it enhances reproduction. Being large, in fact, is usually beneficial, not only in humans but also in many other species. Larger babies survive better during early postnatal life. During childhood, a larger size allows for more successful competition for food. Large size at adolescence is related to earlier sexual maturation. Large adult size is also important. In humans, taller men have greater reproductive success (Pawlowski, Dunbar, and Lipowicz 2000), and larger women have babies of larger birth weight (Kirchengast and Hartmann 1998), who thus have a better rate of survival (McCormick 1985), which in turn may contribute to the greater reproductive success of mothers as well.

What do we know so far? It seems clear that nutritional conditions in utero influence the size, metabolism, and physiology of the fetus. It is also clear that such programming may expose the individual to a higher risk of several metabolic diseases much later in life, especially if there is a mismatch between conditions during pregnancy and the postnatal period (Gluckman and Hanson 2005). When poor conditions are experienced in utero and greatly improved conditions occur in later life, the mismatch results in detrimental health problems. It is less clear whether the thrifty phenotype, which results from developing under poor in utero conditions, should be considered adaptive. A life-history perspective that incorporates the logic behind the theory of mother-offspring conflict suggests that programming benefits the mother more than it does her offspring.

Is reproductive physiology programmed by fetal nutrition as well? If so, what kind of modifications should be expected? Can we postulate that in utero programming of reproductive function is evolutionarily adaptive—that it increases the reproductive success of the individual? These questions have yet to be addressed in the ongoing debate, and they should be considered here.

Fetal Conditions and Adult Reproductive Physiology

When it comes to female reproductive physiology, permanent modifications that cause the female reproductive physiology to become "thrifty" should not be considered adaptive. In favorable environmental conditions, adult ovarian function should be set in such a way that the ability to conceive is always high. A woman with sufficient dietary intake and little physical activity has a lot of energy that can be used to produce a great number of large, healthy offspring. Programmed thriftiness of ovarian function resulting from the mother herself developing in poor fetal conditions would not benefit her fitness later as a well-fed mother.

Can a mother living in a poor environment benefit from having a child with programmed reproductive physiology? Her daughter would always have a lower level of ovarian hormones than the daughters of women who developed in good fetal conditions. I should emphasize here that the Wells hypothesis that programmed physiology of the offspring may be beneficial for its mother refers to nonreproductive physiology. Programmed thriftiness of ovarian function would not benefit either the fitness of the mother nor the fitness of the programmed offspring. There are no conflicting interests when it comes to the ovarian function of offspring. Mothers should want a female child with high reproductive potential, and for females a high reproductive potential is also beneficial. Does this mean that the development of ovarian function would be shielded from the effects of poor fetal conditions? This seems to be the case, but only to some extent. Instead of permanently reduced ovarian function, we observe different kinds of physiological adjustments in response to poor nutritional conditions in utero.

Is Reproductive Function Programmed in Utero?

Only a few studies have investigated the relationship between fetal growth retardation and reproductive physiology in humans. In human males, small birth size correlates with reduced testicular volume, lower testosterone levels (Cicognani et al. 2002), and low adult fertility (Francois et al. 1997). In human females, reduced fetal growth has been associated with impaired ovarian development (de Bruin et al. 1998), reduced size of the ovary and the uterus, and anovulation in adolescent girls (Ibanez et al. 2000; Ibanez et al. 2002).

These changes, which appear to be pathological and deleterious to reproductive fitness, are associated with very low birth weights. Low birth weight—in a clinical context, a birth weight below 2,500 grams—is most likely an evolutionary novelty. Babies of such small sizes had a reduced chance of survival during human evolutionary history. Therefore, the physiological changes observed in such very small babies today should not be used as examples when considering the adaptiveness of permanently changed physiology and metabolism in response to fetal conditions. Natural selection does not detect variation in physiological responses unless the individuals survive long enough to reproduce.

From an evolutionary perspective, much more relevant than pathological reproductive physiology of very small or premature babies is the question of a relationship between modest nutritional challenges during fetal life and the effects of such challenges on reproductive physiology many years later. My colleagues and I addressed this question in our study of Polish women. In our study, adult women collected saliva samples for hormone measurements for an entire menstrual cycle and provided us with information about their weight and length at birth. The women who had been relatively fat at birth had 22 percent higher levels of estradiol during the menstrual cycle than the women who were born as relatively skinny babies (Jasienska, Ziomkiewicz, Lipson et al. 2006). The ponderal index, calculated as birth weight/birth height^3, was used as a measure of a newborn's fatness and an indicator of its nutritional status. It is not a perfect indicator, but neither is the birth weight of the child because a child who was nutritionally deprived during late gestation could have a "normal" birth weight but be long and skinny. Such a long and skinny child has a low ponderal index.

Why do women who are born with lower fat levels have lower levels of estrogen? Could we understand such a relationship not by examining developmental or physiological mechanisms but by considering evolutionary, adaptive reasons? Low levels of estrogen are related to a reduced probability of conception. Women with lower levels of estrogen are less likely to conceive in a given menstrual cycle (Lipson and Ellison 1996; Lu et al. 1999), so they have a reduced lifetime probability of conceptions. Natural selection should promote high levels of estrogen because there is such a clear, positive relationship between levels of ovarian steroid hormones and fertility. Short-term suppression of the ovarian function makes evolutionary sense (Ellison 2003a, 2003b) because it prevents maternal investment during times when her nutritional condition is poor, but it is difficult to explain a permanent reduction in ovarian hormone levels as being adaptive.

Reduced estradiol levels in women born small do not seem to be an adaptive response to a low-energy environment. Even if poor nutritional status during pregnancy together with poor maternal status before pregnancy and during childhood development are indicators that environmental conditions are likely to be poor in the future, why would a baby girl be born with changes in reproductive physiology that will result in her ovarian function permanently operating at a lower level? Reduced estradiol levels in women of small birth size seem to be a physiological constraint—the best physiological outcome possible given poor fetal nutritional conditions. Such a conclusion may, however, be too simplistic. Additional analyses on the data from Polish women suggest that we are not dealing with a physiological constraint.

Fetal Conditions and Sensitivity of Ovarian Function

Can we predict what would be an adaptive outcome of fetal programming of ovarian function? What kind of reproductive physiology would be optimal for a woman who, during fetal life, developed her physiology based on intergenerational signals suggesting that environmental conditions are poor? She developed using an "ontogenetic assumption" that energetic conditions during her adult life would be poor as well. But even if such assumptions were correct and conditions indeed are poor, there usually is some variability in environmental quality, given the long human life span. Consistent rainfalls provide better years, good harvests, and reduced physical labor demands. Most beneficial for a woman of reproductive age would be an ability to take advantage of periods of improved energetic conditions with high levels of ovarian function. High levels of ovarian hormones would give her a high chance to conceive during the time when her good nutritional condition allows her to successfully support the energetic costs of reproduction.

She should also be able to respond with suppressed ovarian function to temporary declines in energetic conditions. To accomplish this, her reproductive physiology should have a built-in *high sensitivity* of ovarian response—that is, it should be able to respond promptly to an energy decline. A woman who developed in poor fetal conditions ought to have high sensitivity to poor conditions in adulthood; that high sensitivity should develop based on intergenerational signals that she received during development telling her it is safe to make an assumption that negative changes in her adult energetic condition will be long lasting. If conditions were poor for several past generations, a decline in good conditions that a woman experiences during her adulthood is very likely to persist. In contrast, for a woman who developed in a good fetal nutritional condition and additionally received intergenerational signals in utero of good conditions during the past, a small decline in her own nutritional status in adulthood should not be sufficient to respond with reproductive suppression.

How can we assess the magnitude of sensitivity of ovarian function? What difference should be expected for women going through fetal development in either poor or favorable conditions when they become adults? The difference should not lie in the *overall* level of ovarian activity. When an adult woman is in good shape (from the energetic point of view), she should be able to produce high levels of ovarian hormones. As discussed earlier (Chapter 1), levels of ovarian steroid hormones are affected by a woman's adult lifestyle, and when her energetic status declines she should suppress her ovarian function. One of the factors that has a well-established ability to reduce levels of progesterone and estradiol during menstrual cycles is physical activity.

We may hypothesize that women with low physical activity produce high levels of hormones, regardless of their developmental nutritional experience. When physical activity is very intense, hormone production should be suppressed. This suppression should occur in all women, regardless of their past developmental conditions. However, faced with moderate levels of physical activity, women should show variable responses, and this variation in response should depend on their developmental experience. Women who developed in poor nutritional conditions should respond with ovarian suppression, but women who developed in good conditions should not react to relatively mild energetic stress. Given the positive intergenerational signal the latter received from their mothers during fetal development, a modest deterioration in energetic conditions (in this case, from moderate-intensity physical activity) is an insufficient reason to suppress reproductive function. The intergenerational signal for these women is based on environmental conditions improving shortly, so more intense physical activity is necessary for these women to experience reproductive suppression.

Our study showed exactly these kinds of responses when we compared the estradiol concentrations of Polish women at three levels of habitual daily physical activity (Figure 3.1) (Jasienska, Thune, and Ellison 2006). Based on what we know about physical activity and ovarian function, we would expect the lowest levels of estradiol in women who are the most physically active, moderate levels of estradiol in women with moderate activity, and the highest levels of estradiol in the least active women. In addition, we also compared, at each level of physical activity, the women who were born with either a low or high ponderal index. Levels of estradiol were low in the women who were most physically active, regardless of their size at birth. High levels of estradiol were present in the women with low physical activity, again regardless of birth size. But in women who had comparably moderate levels of activity, significant differences in estradiol levels were observed between two groups of women depending on their size at birth. Women with a small size at birth had reduced estradiol, while women born larger apparently did not suppress estradiol in response to the energy stress of moderate physical activity.

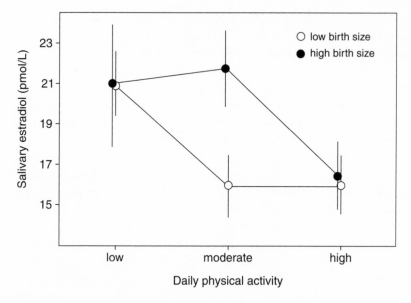

Figure 3.1. The interactive effect of size at birth and level of physical activity in adulthood on mean levels of estradiol in menstrual cycles. Women with low activity have high levels of estradiol, regardless of their size at birth. In women with high levels of activity, the level of estradiol is suppressed, regardless of the woman's size at birth. In women with moderate levels of activity, those who were small at birth have suppressed estradiol levels, but those who were large at birth show no ovarian suppression.

The results of this study fit nicely the hypothesis that a permanent thriftiness of the ovarian function is not programmed in utero because such a permanent reduction in the ability to conceive would clearly be too inflexible: it would lead to reduced lifetime fitness. Undoubtedly, more data are needed, but so far this has been the only study to investigate the relationship of fetal programming to adult energy conditions and its inter-action with ovarian function. The data clearly suggest that ovarian func-tion is programmed in utero, but the programming involves changing the sensitivity of ovarian function, not a permanent reduction in ovarian function.

All this may be relevant for attempting to understand the response of ovarian function in an evolutionary, adaptive context. But is it important for understanding women's health problems and their prevention?

Developmental Programming and Breast Cancer

A lot of evidence points to health problems being related to low birth weight. This has been a serious concern in perinatology and public health for a long time because it is tied to an increased risk of childhood diseases and higher mortality. There is also a relationship between low birth weight and an increase in the risk of chronic adult-onset diseases.

For women, even birth weight is not free from trade-offs. Both low and high birth weights have their disadvantages. A high birth weight may be related to negative consequences later in life. Women born as heavy babies have an increased risk of breast cancer (Ekbom and Trichopoulos 1992; Michels et al. 1996; Ahlgren et al. 2007). One hypothesis proposes that the maternal environment directly affects fetal breast tissue (Trichopoulos 1990). Another hypothesis suggests that the maternal environment influences the development of the fetal neuroendocrine axis (Davies and Norman 2002). Both mechanisms may be important, but the previously described study of Polish women (Jasienska, Thune, and Ellison 2006; Jasienska, Ziomkiewicz, Lipson et al. 2006) provides some support for the latter hypothesis.

The ovarian function of women who had a larger size at birth, as indicated by a higher ponderal index, had lower sensitivity to energetic stress, as shown by the interactive effect of size at birth and physical activity during adult life (Jasienska, Thune, and Ellison 2006; Jasienska, Ziomkiewicz, Lipson et al. 2006). A remarkable relationship exists between the events occurring during fetal life and the production of estradiol in menstrual cycles about thirty years later. Women who were larger at birth, in comparison with the women who were born smaller, had 19 percent higher estradiol levels throughout the entire menstrual cycle, and 25 percent higher levels throughout the luteal phase (Jasienska, Ziomkiewicz, Lipson et al. 2006). Such an increase in estradiol levels, if continuing throughout the entire reproductive span from menarche to menopause, would significantly increase their lifetime estrogen exposure, which is the main contributing factor for an increased risk of breast cancer in women.

Of course, estradiol levels can be modified by changes in one's lifestyle. While, on average, large size at birth is related to higher estradiol levels, this is only the case in women with moderate levels of physical activity during adult life. High levels of activity reduce the levels of estradiol regardless of birth size. These results also clearly suggest that the level of physical activity that a woman needs to reduce her risk of breast cancer depends on her size at birth. Women of larger birth size may need more intense activity to achieve the desired effect of lowering their levels of estrogen.

The determinants of women's health cannot be fully understood without considering the long-term effects of the conditions they experienced during fetal development. In fact, knowledge about conditions during all stages of life—fetal, childhood, and adult—is necessary to provide a comprehensive picture of health. When it comes to reproductive physiology, our knowledge about the impact of fetal and childhood environments on future health status is very limited. We know even less about the interactions that conditions during these different stages of life may have on our physiology.

Two hypotheses related to fetal environment will be discussed in Chapter 4 and Chapter 5. They are based on the assumption that many aspects of development during this stage of life are based on an intergenerational signal that "summarizes" for the fetus information about the quality of life that our recent ancestors lived. An intergenerational signal suggesting that life is good may help explain the low risk of cardiovascular diseases observed in the modern French population: the so-called French paradox. A very different intergenerational experience, resulting from long-term biological deprivation during the years of slavery, may help us understand why contemporary African Americans have low birth weights.

In Chapters 4 and 5, I will continue discussing the potential impact on our biology and health of the intergenerational experience of environmental conditions. I have already looked at some evidence suggesting that the birth weight of a child is determined not only by the nutritional conditions of the mother, but also most likely by the nutritional conditions experienced by female ancestors over the past few generations. Birth weight is an important predictor of health and disease, including the risk of cardiovascular disease; people who had a low birth weight have an increased risk of developing these diseases. Populations differ in their average risk of cardiovascular diseases, and this variation cannot be satisfactorily explained only by differences in the characteristics of their adult lifestyle.

Is it possible that nutritional conditions of past generations have an impact on the risk of cardiovascular disease experienced by people today? In France, the last several generations of women and children benefited from programs that improved their nutritional status and health, and the French today enjoy a relatively low risk of cardiovascular disease. Is there a connection between these two phenomena?

4

Did Child Welfare Reduce the Risk of Heart Disease?

Jane Sharp, a British midwife, was the first woman to write a book for women on pregnancy and childbirth. *The Midwives Book, or the Whole Art of Midwifery Discovered, Directing Childbearing Women How to Behave Themselves in Their Conception, Breeding, Bearing, and Nursing of Children,* was published in 1671. Sharp, with thirty years of professional experience, seemed to have a clear understanding that a good maternal condition is important for a healthy outcome of pregnancy. She listed three requisites for "forming a child": "1. Fruitful seed from both sexes wherein the soul rests with its forming faculty; 2. The mother's blood to nourish it; 3. A good constitution of the matrix to work it to perfection." Thus, the importance of the maternal condition during pregnancy for the health of the newborn baby has been known for a long time. More recently, as discussed in Chapter 3, we have begun to learn about the long-term consequences of the embryonic environment for the biological condition and health of individuals for the rest of their lives.

The French have a diet high in animal fat and, perhaps as a consequence, high levels of blood cholesterol. For an average French person in 1988, animal fat accounted for 25.7 percent of all consumed calories; in the year 1990, an average French man between fifty and seventy years old had a mean total cholesterol serum concentration of 6.1 nanomoles per liter (nmol/L) (Law and Wald 1999). During those same years, the British had very similar fat content in their diet and a similar blood cholesterol level: 27.0 percent of calories came from animal fat, and the blood cholesterol level in British men over fifty years old was 6.2 nmol/L.

If one tried to make a prediction about the prevalence of heart disease in both countries, what would that be? People with any knowledge of risk factors for coronary disease would probably come to the conclusion that both the French and the British face a similar risk of cardiovascular problems—and they would be wrong. In fact, a British man is four times as likely to have heart disease than a Frenchman. A British woman is, astonishingly, six times more likely to have heart disease than a woman from France. The low risk of heart disease in France, despite their high fat intake and high blood cholesterol levels, has been called the "French paradox," and so far no one has been able to offer a convincing explanation.

Some studies show that wine drinking has a protective effect for cardiovascular problems (in fact, any alcohol seems to have a similar effect), and wine consumption is very high in France (Criqui and Ringel 1994). The protective effect of wine disappears, however, during multivariate statistical analyses in which other factors are taken into account. In other words, the difference in wine drinking between the French and the British does not explain the differences in the prevalence of heart disease between these two populations. Another interesting hypothesis claimed that the "time lag" effect is responsible for the observed differences in the prevalence of heart disease between the French and the British.

Malcolm Law and Nicholas Wald (1999) argued that the consumption of animal fat increased in France only relatively recently, while in Britain it was already high several decades ago. Because heart disease results from long-term exposure to risk factors (including an unhealthy diet), there simply has not been enough time in France for high rates of heart disease to develop as a result of dietary changes. An obvious prediction of the "time lag" hypothesis is that the prevalence of heart disease will increase in France in the near future.

It is also possible that other aspects of the French diet or lifestyle in general (for example, long meals, lack of snacking, or use of olive oil) contribute to the low incidence of heart disease in that country. But in this chapter we will examine another intriguing hypothesis to explain the French paradox. David Barker (1999), a British epidemiologist and a pioneer of research on the relationship between the fetal environment and adult-onset diseases, believes that the French have low rates of heart disease because for the last several generations the French did not experience any episodes of major undernutrition during uterine development.

Barker and many other researchers have documented the relationship between fetal developmental conditions and an increased risk of several chronic diseases, including heart disease, many years later. A poor fetal environment may cause permanent changes in the physiology and metabolism of individuals that subsequently may lead to development of heart disease. But is the history of improvement in maternal and child-hood conditions in France so exceptional that it can shed light on the French paradox?

Child Welfare Programs in France

Ironically, maternal and infant welfare movements in the nineteenth and beginning of the twentieth century began in many countries as a result of wars (Dwork 1987, 208–211; Fildes, Marks, and Marland 1992, 1). Wars decreased population size and often resulted in low fertility. In France, as early as the mid-1800s, the government was concerned about "depopulation," which in reality meant concern about depletion of the number of young men available for the French army. The government performed calculations to predict how many extra regiments the army might have if infant mortality was lower.

One of the reasons for high infant mortality in France was a practice known as wet-nursing, where newborn babies were sent from the cities to villages to be nursed by local women. Most of these women had their own babies to nurse, and they often accepted more than one baby to nurse, thus dividing their breast milk between at least two infants and often more. As a result, many wet-nursed urban babies were undernour-ished, in addition to growing up in poor hygienic conditions. After the 1873 legislation that created a national system for the regulation and inspection of wet nurses, the condition of wet-nursed children improved, and the practice of wet-nursing persisted until World War I.

However, infant mortality remained very high. After World War II, France had an infant mortality rate of 110 per 1,000 live births. By comparison, the United States had a rate of 39.8 per 1,000 during the same period. Remarkably, just fifteen years after the war, the infant mortality rates in France were almost the same as those of the United States: 27.4 per 1,000 live births in France versus 26.0 per 1,000 in United States (Bergmann 1986, 73).

France has a remarkably long history of public childcare. Nursery schools *(école maternelle)* were established under religious auspices before the Revolution of 1789 to help poor working parents. Public nurseries, entirely free of charge regardless of the parents' income, were established by the 1820s, and their number grew rapidly. In 1881, a law guaranteed free public primary education to all children, and the *école maternelle* became an integral part of the program (Bergmann 1986, 28–

35). Children could be enrolled in nursery school as soon as they were toilet trained, and they remained there until they were ready for the first grade of school.

In 1945, just after the Germans left France, the French government established a program called *Protection Maternelle et Infantile* (PMI), described as a national program of preventive care for pregnant women and infants. The PMI kept records of all pregnancies and babies, provided assistance for women and children, and identified pregnancies at risk. The PMI's mission was to reduce the impact of inequalities of income and social status on the health of children (Bergmann 1986, 70, 74–85).

More recently, pregnant women, beginning with the third month of pregnancy, receive each month a New Baby Allowance *(Allocation de Jeune Enfant)*. All women, regardless of income or the number of children they already have, receive a fixed payment until the child is three months old. (In 1991, the payment was 875 francs, which was equivalent to US$134.) If a family does not have a high enough income, the payments may continue until the child is three years old. In 2003, the French prime minister announced that a bonus of 800 Euro would be awarded to every mother for the birth of a new baby (Dorozynski 2003). However, a woman can lose her New Baby Allowance if she does not follow a schedule of compulsory, free medical examinations for herself and later free medical checkups for the baby (Bergmann 1986, 59–61).

Only 4 percent of pregnancies in France are not registered during the first trimester. All women who are considered to be at risk (for example, because of young age or having an unemployed husband or partner) are monitored through regular home visits. Women who previously had a baby with a low birth weight are scheduled for frequent medical visits (Bergmann 1986, 79).

Targeting Babies: Breastfeeding and Safe Milk for Weaned Babies

Between 1877 and 1886, infant mortality in France was 226 per 1,000 births, while in England the corresponding figure was much lower at 167 infant deaths per 1,000 births (Dwork 1987, 94). The single most important cause of such high infant mortality in France, as in many other countries, was epidemic diarrhea, which was especially frequent in bottle-fed infants.

Two men, independently, established successful programs aimed at improving infant survival. Pierre Budin worked in Paris from 1892 (Dwork 1987, 98–99), and Dr. Leon Dufour worked in Normandy from 1894 (Dwork 1987, 101–103). Both opened baby welfare clinics that promoted breastfeeding. Although both clinics emphasized the importance of exclusive, prolonged breastfeeding, the mothers who were not able to breastfeed received, almost free of charge, a daily supply of sterilized milk. In Dr. Dufour's clinic:

> Every day the mother received one basket with as many bottles of modified ("humanized" [cow's milk diluted with water and usually with some sugar and cream added to make its composition more similar to human milk]) sterilized milk as the child required for a twenty-four-hour period; the bottles were filled proportionally. Upon the return of the basket-and-bottle set the following day, the mother was given the alternate kit. (Dwork 1987, 102)

In addition to promoting breastfeeding and the use of sterilized milk, the clinics provided continuous medical care for babies during their first years of life.

These two small "natural" experiments achieved spectacular results. In 1898, Budin's clinic took care of fifty-three infants: nineteen were breast-fed, and thirty-four were bottle-fed. The mortality rate among these babies during their first year of life was zero. Infants in the town of Fecamp, where Dufour's Goutte de Lait clinic was located, suffered over 18 percent mortality from diarrhea in the years 1895 to 1896, but mortality was reduced to less than 7 percent in the infants whose mothers attended the clinic.

The lower mortality in the cohort of Goutte de Lait "graduates" continued during the subsequent years of their lives. In comparing the Goutte de Lait infants with the entire infant population of Fecamp, the mortality rates were 4 percent versus 9.5 percent in 1896 to 1897, 2.3 percent versus 12.0 percent in 1897 to 1898, and 1.3 percent versus 9.7 percent in 1898 to 1899, respectively (Dwork 1987, 102). Budin's and Dufour's clinics became more popular with time, and they attracted an increasing number of mothers with infants. In 1898, Dufour's clinic had only 174 infants; three years later, 1,438 babies were under its regular care.

Programs designed to provide a safe milk supply for infants were established not only in France but also in other countries. For example, in England, the first program, the St. Helens milk depot, opened in 1899.

Although these clinics were based on the successful French model, there were important differences. British programs did not emphasize breast-feeding or provide medical care to the infants who received sterilized milk. In addition, the price of milk in France was "graduated according to the ability of the parents to pay" (Harris 1900), but in England one uniform rate was charged; poor mothers in Britain paid twice as much as poor mothers in France. In general, the milk depot idea that successfully worked for many years in France never got fully established in England. The few that existed were too expensive for poor parents, so they reached only a limited number of children.

Tadeusz "Boy" Zelenski, a young Polish pediatrician, became convinced in 1905, after two trips to Paris, that the French model of infant care, especially safe milk for infants who could not be breastfed, should be implemented in his native Krakow (Sterkowicz 1952a). At that time, that area of Poland was under Austro-Hungarian administration, and government sponsorship of such a program was out of the question. Fortunately, a lecture in Krakow by Dr. Zelenski on the topic of infant care proved convincing to a rich countess, Teresa Lubomirska. Lubomirska and her husband had been unable to have children for several years after they married, so she had taken a religious vow to give money to a philanthropic organization in hopes of conceiving a child as a reward. She became interested in Zelenski's ideas and promised to partially sponsor the "safe milk" program.

Doctor Zelenski's institution, named Kropla Mleka (A Drop of Milk), occupied two rooms in a building on the main square of Krakow, and it became operational in July 1905. Cow's milk from the farm of another aristocrat, Count Potocki, arrived each morning, whereupon it was diluted with water in proportions appropriate for infants of various ages, supplemented with milk sugar, transferred to small bottles, pasteurized, and kept on ice until the mothers came to collect it for their children. The milk was not free unless a mother could not afford it, but it was priced in such a way that the institution made no profit.

Providing milk was not the only service offered by this institution. For one hour each day, Dr. Zelenski was available for mothers and children. In return for getting milk, the mothers were required to bring in their infants once per week to have their weight taken and health status evaluated. They also heard advice about the appropriate care for their infants. These services were free of charge for all participating women. Kropla Mleka in Krakow was planned to serve only fifty infants per day, but after just a few weeks it became so popular that twice as many received milk daily. Because the majority of mothers came from Kazimierz, Krakow's Jewish district, and had difficulty traveling every day to the main square, Dr. Zelenski established another Kropla Mleka more conveniently located for the Jewish population at a pharmacy in Kazimierz.

One year after the Kropla Mleka was established, Dr. Zelenski opened another institution with a supplementary function. Again following the successful French model, *Consultations de Nourissons* provided advice to mothers and especially propagated the benefits of breastfeeding. Financial prizes were given to mothers who regularly visited the institution and carefully followed the prescribed methods of care for infants.

Unfortunately, in 1907 Countess Lubomirska, after giving birth to a healthy baby boy, lost interest in philanthropy, and the money stopped (Sterkowicz 1952b). At around the same time, Dr. Zelenski, after learning that his chances for tenure at the Jagiellonian University Pediatric Clinic were slim, left his position and decided to end his pediatric career. Kropla Mleka functioned for several more months under different leadership, but without sufficient financial support and without Zelenski's enthusiasm, it closed in early 1908.

Although many countries had programs of maternal and infant care, often modeled after French examples, in France they developed early, were constantly improved, and provided the most comprehensive care. By providing sterilized milk, promoting breastfeeding, and offering medical care for babies, the French programs resulted in a substantial reduction in the infant mortality rate, the purpose for which they were originally designed. Babies that are breastfed or drink sterilized milk have lower rates of diarrhea and grow more rapidly. Such programs, therefore, resulted in a higher proportion of children with normal growth patterns and good health status, which is important when considering the importance of intergenerational effects of nutrition and energetic conditions on the physiology and health of the population.

Targeting Mothers: Meals for Pregnant and Nursing Women

France was also a pioneer in providing meals for pregnant and nursing women (Dwork 1987, 167–184). In Paris, in 1904, an organization devoted to the promotion of breastfeeding (L'Oeuvre du Lait Maternel) also provided free meals for nursing mothers. One year later, five restaurants in Paris provided such meals, the costs of which were covered by the government and charitable organizations. Two meals were served every day, and mothers could also apply to have additional food served in the morning. Only meat was served in limited quantities. An advertisement for this program stated that "any mother is welcome to come in. She will have to give neither name nor address nor reference of any kind. She has but to show that she feeds her baby" (Dwork 1987, 146). England was quick to follow suit: the first restaurant to offer food to nursing mothers was opened in Chelsea in 1906; although it also included women in the last three months of pregnancy, only one meal per day was offered, and it was not free, even though the price was very affordable (Dwork 1987, 146–147).

It is difficult to determine what kind of effect such supplementation had on maternal and infant health. Judging from the results of the modern studies on maternal supplementation, especially from the Gambia and Guatemala, it seems likely that it had a limited effect on infant birth weight, but probably it did greatly improve maternal condition. Supplementation during pregnancy or lactation, by providing mothers with additional calories, may improve maternal condition in such a way that women take less time to become pregnant again (Prentice et al. 1983; Prentice et al. 1987; Ceesay 1997). Therefore, programs to feed mothers may have been successful in increasing the birth rate but not necessarily in improving the birth weight of babies or the condition of the infants.

Targeting Older Children: Meals at School

A system of school meals was developed in France in the second half of the nineteenth century after the French Revolution (Dwork 1987, 167–184). Legislation in 1867 established a school fund to support educational expenses for poor children, which was implemented in selected regions. Further, legislation in 1882 already required all towns and villages to have such funds available. In part, these school funds were used to provide meals at school for all children. For those whose families were not able to afford the cost, the meals were free.

A commissioner from the British medical journal *The Lancet* who visited French schools to evaluate the school meals system wrote, "Is there not in all this the making of a healthy people and a strong race? Can we in England afford to ignore the example here given?" He added, "Therefore, and without hesitation, if we would preserve the race, without the loss of a moment and without making a single exception, every hungry child must be fed" (Anonymous 1904). Similar programs followed in England but "in a more or less desultory fashion" (Dwork 1987, 176). In 1889, the London School Board revealed that food supplied to school children was inadequate because they received no more than one or two meals per week.

The importance of regular school meals for children's health was evaluated in 1907 in a three-month "experiment" in Bradford, an industrial town in the north of England. Forty children from poor families were selected to receive two meals per day, five days per week. The supplemented children gained 1.2 kilograms, while the control group (the unsupplemented children) gained, on the average, 0.4 kilograms during those three months. The program was run by Ralph Crowley, a school medical officer in Bradford, who, when writing about the success of school meals, noted that for the unsupplemented children "the average gain per year of children of this class and size is not more than 2 kilos . . . for the whole year" (Dwork 1987, 182–183).

In light of data showing a strong positive correlation between maternal rate of growth, the mother's body size in childhood, and the birth weight of her children, good nutrition during childhood seems especially important when considering multigenerational effects on the birth weight and physiology of the developing fetus. In France, the combination of breastfeeding promotion, safe milk, low rates of diarrhea, and school meals contributed to an improvement of the general health status of the population beginning with the establishment of nursery schools in the late eighteenth century.

Many countries attempted to follow the French model of maternal and childhood welfare but with mixed success. Medical practitioners and social workers active in London between 1870 and 1930 observed that malnutrition and poor health were common among women, and East London mothers living in poor neighborhoods were portrayed as "haggard and worn" (Marks 1992, 48). Susan Pedersen, describing the differences between the French and British family welfare systems, wrote:

> French policies reflect what I have termed a "parental" logic of welfare while British choices exhibited a "male breadwinner" logic since, in the former, some portion of the earnings of all adults was forcibly expended in the support of all children, while in the latter both wage and benefit income was directed disproportionately to

men in the expectation that some would use it to support dependent wives and children. (Pedersen 1993, 413–414)

It is likely then that some British children did not benefit much from family welfare.

In following the logic of "fetal programming," we may conclude that due to many generations of improved nutritional conditions French babies come into the world "physiologically programmed" that their future life conditions will be good. This prediction of future conditions is based on the intergenerational experience of past conditions. In France, thanks to a long history of many programs aimed at improving maternal and child nutrition and health, this experience indicates that life conditions, mostly in terms of availability of metabolic energy, are of good quality. In these circumstances, the fetus develops its physiology "ready" for a nutritionally adequate environment—that is, no physiological and metabolic adjustments prepare the physiology of the fetus for poor conditions.

Most modern French citizens do indeed encounter such good conditions during their entire postnatal lives. Therefore, following the logic of the hypothesis of Gluckman and Hanson (2005, chapter 4), we can say that the French people do not experience any mismatch between the predicted and the encountered environments. This "mismatch hypothesis" suggests that poor in utero conditions lead to an increased risk of metabolic diseases in adults *only* when developmental conditions and adult conditions are different. A lack of such a mismatch in the French population may lead to the low risk of cardiovascular disease despite their high-fat diet and the resulting high blood cholesterol levels.

In Chapter 5, we continue thinking about early developmental conditions and their potential long-term impact on health. Just as good conditions experienced by several past generations may have a positive impact on the health of the contemporary population, long-lasting poor conditions may influence it negatively. In Chapter 5, I talk about the past of African Americans during the period of slavery. This population experienced nutritional deprivation, excessive workload, and a high burden of disease. Today, people living in the United States who are descendants of slaves are born with weights that are significantly lower than that of other Americans. Is it possible that there is some connection between the conditions in the past and the birth weights of today?

5

INTERGENERATIONAL ECHOES OF SLAVERY

Babies of American mothers of African descent have a higher risk of prematurity and low birth weight compared with the babies born to American mothers of European descent. National vital statistics reports for 2003 show that African American infants had an average birth weight of 3,122 grams compared with an average white infant weight of 3,384 grams. Out of all African American infants born that year, 11.58 percent had a clinically low (less than 2,500 grams) birth weight in comparison with 5.11 percent of all white infants. In the same year, the U.S. Hispanic population had the average birth weight of 3,324 grams, and 5.55 percent of infants with low birth weight. Remarkably, the average birth weight and the percentage of low-birth-weight babies did not change for African Americans for many years; a century ago, the African American birth weight was 3,183 grams (Costa 2004).

Note that I use the terms "race" and "racial" out of necessity to connect with the substantial body of older literature in which these terms were prevalent. I should emphasize that dividing people into racial groups is not biologically or genetically justified (Bogin 1999). Thus, in the context of this book I am concerned only with these terms as identifiers of the geographic origins of the analyzed populations.

Clearly, the differences in birth characteristics between African Americans and European Americans have resulted to some extent from differences in socioeconomic status. Education influences income, and both these factors have an impact on the nutritional status of the mother, her access to health care, and her ability to follow preventive care measures during pregnancy. However, even after we take socioeconomic inequalities into account, significant differences in birth characteristics remain. If one puts side by side the white and black populations with comparably low incomes, black women have babies weighing 200 grams less on average than the babies of white women (Goldenberg et al. 1996). Newborns of black mothers are born at significantly younger gestational ages (38.1 weeks versus 38.7 weeks for white mothers), and their mothers have a higher percentage of preterm deliveries (16.7 versus 11.3 for white mothers) and low-birth-weight infants (14.8 versus 8.7 for white mothers). In the study that reported these results, the ethnic groups differed in factors affecting the likelihood of poor birth outcomes—but the black women were better educated, smoked less, scored better on psychological tests, reported having better housing, and had higher body weights than the white women. Out of the numerous tested factors, only maternal height, weight, blood pressure, diabetes, and smoking had any impact on birth outcomes, but none of these factors explained all the differences among the ethnic groups. After adjusting in the statistical analysis for all the potential risk factors, black babies still weighted 139 grams less than white babies. The investigators concluded that only about 30 percent of the variation in birth weight was explained by the examined maternal characteristics, with hypertension being the most important risk factor among black women. A study in South Carolina found that, even after controlling for educational level, use of prenatal care, and the interval since the previous birth, black women still had a higher proportion of low-birth-weight infants (Carlson 1984).

Dora Costa (2004) examined the factors responsible for the difference in birth weight between African Americans and European Americans based on two data sets. The first set came from the Johns Hopkins Hospital in Baltimore from its historical records for the years 1897 to 1935, and the second set came from the 1988 National Maternal and Infant Health Survey (NMIHS). There were important differences between these data sets: the NMIHS was a survey of a random sampling of U.S. births in 1988, and the Johns Hopkins records were for the Baltimore urban population only. Both studies used a variety of socioeconomic and health variables in their analyses. In the period from 1897 to 1935, based on live-birth statistics only, European American babies had a mean birth weight of 3,422 grams while for African Americans it was 3,183 grams, or 239 grams (7 percent) lower. In 1988, the average birth weight of European American babies was 3,426 grams and of African Americans 3,132 grams (294 grams or 8.6 percent lower).

In the 1897 to 1935 data set, the observed fact that European American and African American women differ in their children's birth weight was explained by prematurity and gestational age. Thirty percent of the observed variation in the incidence of prematurity among mothers of European and African descent was explained by differences in the incidence of syphilis infection among them. However, it is important to note that, even after taking into account the different rates in premature birth, the 91 percent interracial variation in full-term birth weights remained unexplained. None of the socioeconomic factors was capable of explaining these differences between these two populations.

In the 1988 NMIHS study, the highest fraction of racial variation in birth weights was explained again by differences in prematurity. The fraction of premature babies was 0.066 for European Americans and 0.172 for African Americans. Differences in marital status, in turn, explained 16 percent of the difference in prematurity, with more African American women being unmarried. Interestingly, for the NMIHS sample it turned out that, irrespective of race, maternal birth weight had a statistically significant impact on the birth weight of the child. An extra 100 grams in maternal birth weight increased the child's birth weight by 16 grams, or by 14 grams when only full-term births were analyzed. However, the different birth weights of the mothers explained only a small fraction (6 to 7 percent) of the racial differences in their children's birth weights. The racial difference in maternal weight gain in pregnancy was also not a factor in explaining the racial variation in birth weight because it contributed to only 4 to 6 percent of that variation.

To summarize, when additional information about maternal birth weights and weight gain during pregnancy was used in the analyses, "only" 70 percent of the racial birth-weight gap was left unexplained in the NMIHS study. It represented a substantial improvement when compared with the 91 percent of racial variation in birth weights left unexplained in the Johns Hopkins study. But these results also show that, even after having carefully taken into account the many differences existing between these two populations, the substantial difference in birth weight still remained a mystery.

Slave Children

African Americans have a long, multigenerational history of nutritional deprivation, excessive workloads, and poor health due to years of slavery and the postslavery period of economic hardship. Although there is a continuing difference of opinion among the researchers about the nutritional status of adult slaves, there is general agreement that slave children indeed experienced very poor nutritional conditions. No measurements of the birth weights of slave children are available, but they have been estimated to have been 2,330 grams on average, based on measurements of children's heights at young ages (Steckel 1986a). Slave children grew more slowly and were shorter than white children during that period. Babies were breastfed, so they probably received relatively good nutrition during this early stage of life, but the quality of their diet dramatically plummeted after weaning (Kiple and Kiple 1977b). Plantation owners believed that a child's diet should consist of fat, cornbread, and hominy— dried corn kernels treated with alkali until the germ is removed (Kiple and Himmelsteib-King 1981, 97). "Clear" (lean, instead of fatty) meat would "debilitate" children and "until puberty meats should not enter too much into the diet," and vegetables also were "not proper for them" (Kiple and Kiple 1977b, 288).

Adult slaves were often lactose intolerant, but young children were still able to digest milk; however, the data show that milk was available only seasonally. Slave owners were aware that a diet high in animal fat was not healthy, but they believed that "negroes and white people are very different in their habits and constitution, and that while fat meat is the life of the negro it is a prolific source of disease and death among the whites" (Kiple and Kiple 1977b, 289). Slave owners were convinced that, for slave children, corn and fat were "peculiarly appropriate" (289). Therefore, slave children grew up having a diet high in fat and carbohydrates but deficient in protein, calcium, magnesium, and iron (Kiple and Kiple 1977b).

A poor diet should lead to higher morbidity—a higher rate of diseases. Slave children more often than white children suffered high rates of typical childhood diseases. They frequently became ill with diarrhea, neonatal tetanus, convulsions, diphtheria, respiratory diseases, and whooping cough (Savitt 1978, 120). Mortality rates in early childhood may be used as a sensitive indicator of the health and living standards of a population (Haines 1985), and early data on mortality among the black population is available from the 1850 U.S. census. The mortality of black children from their time of birth through the next nine years of life was twice as high as that of white children of the same age (26.3 versus 12.9 per 1,000, respectively; Kiple and Kiple 1977b, 290).

In the life of a child, surviving the first year presents the greatest challenge. For each 1,000 slave children born, 350 did not survive their first year of life, a shockingly high number. Comparable statistics for the general U.S. population show that just 179 infants per 1,000 died during the first year (Steckel 1986b). To put these values in the proper context, please note that the equivalent mortality rate for the United States was 6.1 in 2011. The mortality rate for slave children was also twice as high as the average for the U.S. population in early childhood (for ages one through four: 201 versus 93 per 1,000, respectively) and between ages five and nine (54 versus 28 per 1,000, respectively) (Steckel 1986b).

Slave children began working at very young ages, although considerable variation existed among plantations and among individual slave owners (King 1995, 21–41). Jacob Branch, a former slave, said about his childhood, "Us chillen start to work soon's us could toddle" (King 1995, 23). Others remembered that they had to work in the fields before they were big enough to use tools of regular size (King 1995, 23). Young children often worked taking care of babies, and after the age of ten they worked at regular domestic, agricultural, or industrial jobs (King 1995, 25). Forty-eight percent of children began working before the age of seven, and 84 percent before they reached eleven. Girls began working at even younger ages than boys, and they were more productive at the same tasks, such as picking cotton (Steckel 1996, 44). Usually, once children started working they were entitled to additional food. The planter McDonald Furman said that "full allowances" should be received by "small [children] that work in the field," but children who did not work got only "half allowance" (King 1995, 22).

Records of slave body heights bear the unmistakable imprint of the poor nutritional and health conditions they experienced. Data on the heights of individual slaves are available, thanks to the 1807 Congress bill for the abolition of the slave trade (Steckel 1986a). Coastwise trade of American slaves was then permitted, but the law did not allow the African slave trade. Captains of ships that transported slaves between American coasts were required to keep records of their ages and heights to provide proof that the slaves arriving in American ports were not being illegally transported from Africa (Steckel 1986a). These records show that by the age of three, slave children were at the 0.2 percentile of modern height standards, which means that 98 percent of today's three-year olds are taller than an average slave child. This statistic, according to Richard Steckel (1986a), suggests that the standards of living were lower for slave children than for children from the contemporary slums of the poorest African countries!

Poor nutritional status of adolescent girls can further deteriorate once they become pregnant. An adult woman needs energy only to support her basal metabolism and physical activity, but an adolescent must, in addition, allocate energy for support of her own growth processes (Wallace et al. 2004). An adolescent mother must allocate energy to three major functions—fetal growth and development, her own metabolism, and her own growth—which creates a substantial bioenergetic challenge.

To understand the nutritional problems faced by slave mothers, one needs to know the age at first birth for slave women. Historians Robert Fogel and Stanley Engerman (1974, 137–138) have estimated a mean age at first birth of 22.5 years and a median age of 20.8 years, with 40 percent of women giving birth before they were twenty years old. These numbers have been criticized on the grounds of being based only on live births and the number of children who were alive at the time the demographic survey of slaves was conducted (Cody 1977). Such an approach could result in biased estimates; we may conclude that these estimated means and medians are probably too high.

Indeed, another study examining age at first birth at several plantations showed that it varied from 17.7 years of age to 19.6 years (Gutman 1976, 50 and 171; Gutman and Sutch 1976, 142–144). Based on the Ball family plantation records, the mean age at first birth calculated for five periods ranging from 1750 to 1839 ranged from 19.1 to 20.0 with no significant changes with time (Cody 1977). These data suggest that some slave women did not face competing energy demands from their own growth while pregnant, but more telling are the data showing the distribution of maternal age at first birth. Between 10 and 20 percent of all women had their first child between the ages of twelve and sixteen (although the sample size was very small for the period when 20 percent of women gave birth). More than 50 percent of all women gave birth to their first child between their seventeenth and twenty-first year. Taken together, these data suggest that many women were still growing themselves while pregnant or lactating with their first and possibly also their second child.

Women seem to be more sensitive to adverse economic conditions than men (although it should be noted that males have higher mortality risks at most ages). In the United States during the early decades of the twentieth century, the nutritional status of men significantly improved, as indicated by an increase in stature, but no comparable change occurred in women (Wu 1994). Heights of women in nineteenth-century Scotland and Ireland during the industrial revolution showed a steady, decreasing trend for cohorts born between 1800 and 1840 (Riggs 1994). During the same period, the heights of men remained largely unaffected, except for an actual increase in heights for those born in 1840. It is unclear if access to resources changed differentially for both genders, or whether inevitable trade-offs between growth and reproduction were responsible for the decrease in stature for women. Perhaps, when conditions improve, after a longer period of nutritional hardship, additional resources are allocated in women to earlier reproduction and not to their own growth.

Nutrition and Workload of Adult Slaves

It is difficult to reliably assess the average nutritional status of the African American population beginning with years of slavery because of the substantial variation in lifestyle imposed by different types of plantations. Slaves from the Lower South were shorter than those from the Upper South, as the rice culture of South Carolina and Georgia demanded more physical labor than the tobacco culture of Virginia and Maryland (Komlos 1994). An assessment of nutritional status must take into account not only caloric intake but also energy expenditure. People involved in very intense physical labor, as most slaves were, cannot achieve good nutritional status on the same diet as a sedentary person. Nutritional status can also be estimated indirectly by examining body height, especially its secular trend. The body height of children at various ages and their rates of growth are also very useful as indicators of nutritional status. However, one needs to remember that stature and rate of growth are not only influenced by a net difference between energy intake and energy expenditure, but also by other factors such as infections and, of course, genetics.

Historians have presented conflicting views on slave nutrition. Robert Fogel and Stanley Engerman in their classic book published almost forty years ago presented an argument that slave diets were energetically and nutritionally adequate (1974, 107–144). Based on an analysis of census documents of cotton plantations, they estimated that an adult male slave diet provided 4,185 kilocalories per day. The diet of slaves on rice plantations was estimated to yield between 3,150 and 4,200 kilocalories for an adult male (Swan 1972, 252–254). An intake of 4,000 kilocalories per day is high and adequate for most physical occupations, but such high estimates were criticized for being calculated from caloric and nutritional content of raw rather than processed foods (Blonigen 2004). Processing causes food to lose volume, energy content, and nutrients. Beth Blonigen (2004) estimated both energy intake and expenditure for adult male slaves on rice plantations in Georgia and South Carolina in the 1860s. A typical diet yielded 3,162 kilocalories per day when food was analyzed in its raw, unprocessed state, but after adjustment for losses from food processing, the average caloric intake of an adult male slave was reduced to only 2,856 kilocalories per day (Blonigen 2004, 18).

The daily energy expenditure of adult males depended on the season and has been estimated to range from 4,400 to 8,700 kilocalories during the harvest season when slaves worked for about fifteen hours per day harvesting rice (Blonigen 2004, 12). Such estimates of energy expenditure seem extremely high. In the very energetically demanding Tour de France cycling race, energy expenditure ranges between 7,020 and 8,600 kilocalories per day, which is probably the highest level of prolonged energy expenditure ever reported for humans (Saris et al. 1989). In comparison, male army soldiers expend between 3,230 to 5,020 kilocalories per day, and Antarctic explorers from 4,420 to 5,500 kilocalories per day (Peterson, Nagy, Diamond 1990). The assumption that all fifteen hours of work performed by a slave laborer would be heavy labor is unrealistic. In contemporary agricultural populations with demanding labor, such as Guatemalan sugarcane cutters, energy expenditure ranges from 2,579 to 4,086 kilocalories per day (Immink 1979). Eskimo hunters expend from 3,310 to 4,350 kilocalories per day (Shephard 1980).

More realistic estimates of the daily energy expenditure of male slaves can be derived from the Louisiana sugar plantations during cane-planting season. The mean energy expenditure was estimated at over 5,100 kilocalories per day, with an energy intake that was 20 to 40 percent less than the expenditure (Follett 2003). During the harvest season, with workdays lasting between twelve and sixteen hours, the intensity of labor was on the verge of physical capacity: "the fatigue is so great that nothing but the severest application of the lash can stimulate the human frame to endure it" (Thomas Hamilton, former slave, cited by Follett 2003). During the period of most intense labor, slaves on sugar plantations were allowed to drink sugar cane juice (dissolved sugar) to supplement their diet with additional calories.

Extremely high levels of energy expenditure cannot be easily compensated for by intake, even when additional food is freely available. The evidence comes mainly from nonhuman species, but it is well established that there are physiological limits to the rate of food conversion into usable energy (Saris et al. 1989; Peterson, Nagy, Diamond 1990; Weiner 1992; Konarzewski and Diamond 1994; Suarez 1996). When an individual is experiencing high energy demands, increased food intake, at some point, can no longer supply that individual with more energy.

Metabolism that is completely fueled by energy intake without the need to use energy reserves accumulated by the body is called *sustained* (Peterson, Nagy, Diamond 1990). Sustainable energy budgets are thought to be limited by the maximum rate of energy assimilation (Weiner 1989; Weiner 1992; Hammond and Diamond 1994), which in turn is constrained by one or two factors. The first constraint is set by the physiological capacity to digest food; the second results from the limited capacity to absorb nutrients and energy from already digested food (Weiner 1992). Tour de France cyclists during competition consume a special, energy-dense diet that allows a high rate of energy assimilation. But even given this superb diet, these cyclists probably reach the upper limits of sustainable energy budgets—that is, they would not be able to generate more energy with an additional increase in food consumption (Peterson, Nagy, and Diamond 1990). The diet of slaves most certainly was not characterized by comparable energy density and had a much lower rate of energy assimilation. For hard-working slaves, high levels of energy expenditure had to be compensated for by energy generated from body fat reserves.

Most ex-slaves interviewed in the Slave Narrative Collection, which comprises more than 2,000 interviews with former slaves living in seventeen states during the years 1936 to 1938, believed that they were well fed in slavery (Joyner 1971), but Joyner (1971) cautions that a potential bias may be involved: the people who experienced harsh conditions had very little to report about food. A North Carolina slave said, "Master made us all eat all we could hold" (Yetman 1999, 97), while another ex-slave reported, "The white folks fed us. They give us as much as they thought we ought to have" (304). Some slaves, however, remembered being hungry: "They didn't half feed us either. They fed the animals better. They gives the mules roughage and such, to chaw on all night. But they didn't give us nothin' to chaw on" (116).

Even if the slaves' diet was sufficient in calories, as in the controversial view of Fogel and Engerman (1974), it was surely deficient in many crucial nutrients (Kiple and Kiple 1977a). Kenneth and Virginia Kiple (1977a) argue that, given a slave diet of cornmeal and pork fat, which was deficient in niacin and protein, slaves must have had high rates of a nutritional disease called pellagra. Pellagra is common in populations with a high consumption of niacin-lacking corn. Niacin can be obtained by conversion of tryptophan from milk, but the U.S. population of African descent is characterized by a high frequency (about 80 percent) of lactose intolerance. Lactose-intolerant people cannot metabolize lactose (milk sugar) and thus cannot drink milk without severe intestinal discomfort. Due to their inability to digest milk, the slaves' diet was also low in calcium, as the intake of this mineral from green vegetables was only seasonal. It is likely that their consumption of iron was also too low, especially by women and children (Kiple and Kiple 1977b), as most protein rations given to slave families were consumed by the adult men (Morgan 2008).

On plantations, food was usually distributed once per week to adult slaves or to a representative of a slave family (Gibbs et al. 1980). Food provided by plantation owners was in many cases supplemented by the slaves themselves from their own gardens, and from fishing or hunting. As one ex-slave recalled, "Mighty lot of fun to catch them fishes but more fun when they is all fried brown and ready for to eat with a piece of hot pone" (a type of cornbread) (Joyner 1971). On some plantations, however, owners did not allow slaves to hunt or fish. "As for fishin', we never did none, 'cause we hadda work too hard. We worked from can to can't" (Yetman 1999, 56).

There are very few estimates of the energy intake or expenditure of female slaves. Some studies have documented the division of labor between male and female slaves as regards those working in the fields and those working as domestic helpers. In the fields, most activities with very few exceptions were performed by both genders (Gibbs et al. 1980). Only log rolling and corn shucking were listed as activities not commonly performed by women. In domestic tasks, men only worked serving food or as personal servants, whereas women did these and all other activities, such as cooking, cleaning, washing clothes, and sewing. Based on data using known tasks assigned to female slaves, Gibbs and coauthors (1980) calculated the following rates of energy expenditure for tasks performed for five hours: heavy tasks resulted in energy expenditure from 1,800 to 3,000 kilocalories, moderate tasks from 1,200 to 2,770 kilocalories, and light tasks from 360 to 1,170 kilocalories. A hypothetical man doing five hours of heavy labor, five hours of moderate labor, six hours of light labor, and eight hours of sleep would have a daily energy expenditure of 6,306 kilocalories per day. A similarly working woman would expend less energy due to a lower body weight, but her daily expenditure would still be very high.

Work and Pregnancy

Some plantations managers expected pregnant women to work as hard as before the pregnancy: "They forced women by threats of violence— one promised 'to break her belly with foe-foe pounder'—a stout wooden stick used to process plantain—to continue their regular work routine and subjected them with sadistic zeal to standard punishments" (Turner 2002). Most plantation manuals and daily work records show that slave women were usually permitted to work less intensely during more advanced pregnancy, but they were not allowed to reduce their workload before the fifth month of pregnancy (Campbell 1984). Cotton plantations' archives document that nonpregnant women collected, on average, 87.8 pounds of cotton per day; during the period of one to four weeks before giving birth, women worked at 75 percent of their previous intensity. Their output decreased at that time to 67.0 pounds of cotton collected per day, and decreased further during the week of giving birth and a week after, when women collected just 31.3 pounds (Steckel 1986b). One could comment that collecting 14 kilograms of cotton in the days immediately before and after childbirth is no small task.

High energy demands placed on slave women are suggested by the documented significant seasonality of births. Seasonality of births implies seasonality of conceptions; conception itself, while clearly affected by many factors, depends on ovarian function. Ovarian function, as discussed earlier (Chapter 2), is very sensitive to increases in physical activity and reacts with suppression. On the Ball plantation during 1735 and 1865, most births occurred during August, September, and October, which means most conceptions occurred during the winter months (Cody 1977) when there was lower work intensity. Conceptions were least likely to occur between May and July, at times of highest labor demand (for example, when cotton was ready for harvest). The frequency of conceptions rose in August, which may indicate an improvement in the nutritional status of women from a postharvest increase in food supply. As the rates of conception still rose after August, this suggests that the reduction in energy expenditure may have been more pronounced, indicating a lasting effect in relaxing ovarian suppression rather than improved diet.

Plantation owners noticed themselves a relationship between work and fertility in slave women. In 1774, a plantation owner from West India wrote, "Those Negroes breed the best, whose labour is least, or easiest. Thus the domestic Negroes have more children, in proportion, than those on penns; and the latter, than those who are employed on sugar-plantations" (Morgan 2008, 238). "Penns," in this context, were live-stock pens, where the work did not involve as much backbreaking labor as the work in the sugar fields.

Intense physical labor causes ovarian suppression, but is it also related to the outcomes of pregnancy once a woman conceives? The effect of physical activity in pregnancy has most intensely been studied when the physical activity is recreational exercise. Much less frequent studies of women who perform intense physical labor as occupational work are often confounded by the poor nutritional status of such women. Nevertheless, these studies have indicated that in women from developing countries there usually is an association between heavy physical labor and lower birth weights or shorter gestations. Ethiopian mothers who performed intense physical work during pregnancy gave birth to babies who weighed on average 210 grams less than the babies of the mothers with lower activity (Tafari, Naeye, and Gobezie 1980). In pregnant British women, daily physical activity that included physical work did not influence birth weight (Both et al. 2010). The methodology used in the latter study did not allow for measuring physical activity, so it is unlikely that the physical activity of these women was very intense.

For pregnant women, intense physical activity creates conflicting physiological demands. Distribution of blood flow in the uterus is changed by physical activity with potentially harmful consequences for the fetus such as hypoxia, hyperthermia, and a decreased intake of carbohydrates (Sternfeld 1997). What follows is a reduced rate of intrauterine growth and a lowering of the baby's birth weight. There is also the risk of increased uterine contractility directly caused by heavy workloads, which may lead to preterm labor. Although many studies of exercising women do not confirm the negative effects of activity on birth weight (Haakstad and Bo 2011), others show that birth weight is reduced in pregnant women who exercise (Chasan-Taber et al. 2007). Australian mothers who did more than four 30-minute sessions of vigorous exercise per week had babies whose average birth weight was as much as 315 grams lower compared with the babies of women who did not exercise (Bell, Palma, and Lumley 1995).

In another study, women who were physically active before becoming pregnant were randomly assigned to one of three exercise regimes during pregnancy (Clapp 2000). All women exercised until the twentieth week of pregnancy, doing between twenty to sixty minutes of exercise five times per week. Past the twentieth week, women from the first group increased their duration of daily exercise, women from the second group maintained the same daily duration, and women from the third group reduced their daily exercise duration. Women from the first group had smaller babies and a more than 20 percent lower placental volume in comparison with women from the third group. The babies of the first group were 45 grams lighter, and their body fat was reduced by 30 percent. It is important to note that in this study all women followed a quite intense exercise regime, and the described differences in birth outcome were between women involved in exercise of different intensity, not between exercising and nonexercising women. The type of exercise may also be important, and it is likely that weight-bearing exercise is necessary to have a pronounced effect on birth weight (Clapp 2000).

In general, a moderate amount of exercise during pregnancy is not detrimental for healthy, well-nourished women. However, intense physical work, at the level that slave women experienced, especially during the harvest season, might have led to a pronounced reduction in the birth weight of their children.

Heavy physical work by future mothers during pregnancy may also affect the mortality of their babies. The seasonal pattern of workload on cotton plantations corresponded to the seasonal variation in infant mortality during the first month of life, which suggests a clear causal relationship: seasonality of maternal workload may have caused this observed pattern of mortality (Steckel 1996, 53–57). We may also derive a similar conclusion from indirect data from the Kollock cotton plantation in Georgia: the women whose infants successfully survived their first year of life had been allowed to spend more days off work during their pregnancy than the women whose children died (27.2 versus 19.2 days, respectively; Campbell 1984).

A comparison of the birth weights of children born to nutritionally deprived women during the Dutch hunger winter of 1944–1945 showed that the energetic status of the mother during the last trimester of pregnancy had the most pronounced effect on reducing newborn size. Women exposed to famine during the last trimester had babies who were almost 300 grams lighter compared with babies born after the famine (Painter et al. 2005; Painter et al. 2007). Similarly, slave women who experienced increased workloads during the last months of pregnancy probably had babies with very low birth weights.

Birth weight is a strong predictor of morbidity and mortality during early life. Therefore, seasonal increases in workloads were most likely

114

responsible for seasonal variations in birth weight and the related variation in infant mortality. Infectious diseases are also capable of reducing birth weight. Mothers with malaria are known to have smaller babies than healthy, uninfected mothers (Akum et al. 2005), and infections with syphilis, yaws, and elephantiasis, which are more likely to occur in malnourished women, increase the likelihood of miscarriages and stillbirths (Morgan 2008).

The fertility of slave women was relatively high, with an estimated seven (Farley 1965) or even eight (Zelnik 1966) children born on average during the reproductive span. Short durations of breastfeeding and high infant mortality contributed to high fertility rates. Breastfeeding in slave women lasted probably for less than a year, and supplementation with poor quality, low-protein pap and gruel began as early as the second or third month of infant life (Steckel 1986b).

In the United States, the conditions of many slaves did not improve after the abolition of slavery, and freed slaves were treated as second-class citizens. Ex-slaves were not employed in skilled jobs, and many lived in a share-cropping system (Robertson 1996, 29). Under the sharecropping system, farmers worked on land they did not own in return for part of the crop production or a wage. "Taxation and fiscal policies were used to transfer income from blacks to whites, perhaps more effectively . . . than had been possible under slavery . . . Time on the cross did not come to an end for American blacks with the downfall of the peculiar institution" (Fogel and Engerman 1974, 171). The 1900 census data showed that African Americans suffered higher mortality than Euro-Americans from all major diseases except cancer (Rose 1989). Skeletal analysis of the postslavery population of southwest Arkansas has suggested that nutritional deprivation continued, both during childhood and during adult life (Rose 1989).

Caribbean Slavery

The United States was not the only country using slave labor in the eighteenth and nineteenth centuries. Approximately 400,000 to 500,000 African slaves were shipped to the United States, and 4 to 5 million were imported to the Caribbean. Living conditions of slaves in the Caribbean were as harsh and, in some aspects, even more difficult than conditions for slaves in the United States. As Higman wrote, "The ultimate test of the well-being of the slave populations of the British Caribbean lay in their capacity to survive" (1984, 303), and surviving was not an easy task. After abolition of the Atlantic slave trade in 1807, no new slaves arrived from Africa, and the slave population began to decline in number. Between 1807 and 1834, the number of slaves declined from about 775,000 to 665,000 (Higman 1984, 3).

Demographic data suggest that Caribbean slaves suffered very high mortality rates and relatively low fertility, with a resulting negative rate of natural increase. The reasons for this are clearly complex, but biological factors such as diet, work, and disease were important. The rate of natural increase varied over time and by the type of slave colony, with the lowest rates reported for sugar colonies. For example, in the sugar colonies on St. Lucia between 1815 and 1819, there were 15.4 registered births per 1,000 individuals and 46.9 registered deaths, which meant a negative rate of natural increase of 31.5 per 1,000 slaves. Other types of colonies had a similar number of births, but some (such as cocoa-producing colonies) had a lower number of deaths—28.7 per 1,000, which still resulted in a negative (but less drastically so) rate of natural increase of 13.2 per 1,000 people (Higman 1984, 326).

Birth rates for Caribbean slaves were lower than for most slaves in the United States. In Jamaica, the birth rate for Creole-born slaves was estimated at about 25 to 27 per 1,000 individuals (this demographic index is called the crude birth rate), while for U.S. slaves it was about 50 to 55 per 1,000 people (Kiple 1984, 112). The U.S. slave women on the most work-demanding sugar plantations had mean interbirth intervals of twenty-five months, whereas U.S. slave women on cotton plantations, where the work, while still intense, had a shorter daily duration and varied seasonally, gave birth to children every 15 to 16 months (Follett 2003). Interbirth intervals of the Caribbean slave women probably resembled more those of the U.S. slave women working on sugar plantations. In the Caribbean, women on coffee plantations gave birth at younger ages than women on sugar plantations (Geggus 1996, 267), suggesting that hard work at sugar production had a suppressive effect on early fecundity, both by delaying the age of menarche and by suppressing ovarian function during early adulthood. A later age at menarche is related to lower levels of ovarian steroids and a lower rate of ovulatory cycles, and therefore lower fecundity for the years after menarche (Apter and Vihko 1983; Vihko and Apter 1984). It is estimated that in the mid-eighteenth century about half the Caribbean female slave population was childless (Morgan 2008).

Sugar plantations in the British Caribbean after 1807 were organized on a task-work system, in which slaves were given a specific task to complete (Higman 1984, 179–180). A task was the amount of work expected to be completed in one day. A typical workday was twelve hours in Jamaica and ten hours in most of the eastern Caribbean (Higman 1984, 187). The estimated total annual number of hours that slaves spent working for their owners was 4,000 for field slaves in Jamaica and 3,200 in Barbados. Rural antebellum (i.e., before 1861) slaves in the United States worked about 3,000 hours per year, while factory workers in Britain during the 1830s worked about 2,900 hours (Higman 1984, 188).

After finishing their plantation work, slaves usually tended crops in their own fields, which provided the core of their diets. By the end of the eighteenth century, it was customary that slaves did not work on Sundays or for two to four days during the Christmas holidays; in some colonies, they were also let off work during other holidays (Higman 1984, 180).

Height, an Indicator of Nutritional Status

Among Caribbean slaves, those born in Africa were shorter than those born into slavery. In a sample from Trinidad, the African-born adult men were 2.6 centimeters and African-born women 3.1 centimeters shorter than slaves born in the Caribbean (Higman 1979). Kenneth Kiple (1984, 24) has suggested that the increase in height of slaves compared with their native populations in Africa was potentially related to the higher protein content in slave diets in comparison with native West African diets. There were also differences in height among the different types of slave colonies. The Caribbean-born slaves from sugar-producing colonies were significantly shorter and had less rapid growth in childhood than the slaves born in the "marginal" colonies (Higman 1979). Marginal colonies grew agricultural crops other than sugar, coffee, and cotton; in addition, these colonies were involved in salt production, logging, and fishing. It is believed that slaves in those colonies had more time to garden and fish and thus had better nutritional status (Higman 1979). People from marginal colonies in the Bahamas were even taller than U.S. slaves.

Historical growth curves show that slaves born in Guyana had similar growth patterns as slaves from the United States until about seven years of age, after which U.S. slaves grew more rapidly. For the U.S. slaves, there are no separate data for the African-born and U.S.-born slaves, but on average they were taller than the Caribbean slaves. From 1826 to 1860, the average height of U.S. male slaves aged twenty-five to forty years was 171.2 centimeters, compared with in Trinidad in 1813 where the average for both African-born and Caribbean-born slaves was only 164.5 centimeters. The comparable figures for women were 159 and 154.4 centimeters, respectively (Higman 1979, 1984).

Higman (1979) has argued that the differences in height observed between the U.S. and Caribbean slaves were the result of differences in caloric intake. Slaves in the United States received a daily ration of 230 grams of salt meat and 900 grams of corn, but Trinidad slaves had only 195 grams of meat and approximately 520 grams of corn. Even so, U.S. slaves were shorter than white Americans. Southern male farmers were on average 3.8 centimeters taller than the slaves born in a similar location during a similar time period (Margo and Steckel 1983).

The exact nutritional and caloric content of Caribbean slave diets is difficult to estimate, but some sources suggest that the average plantation food allowance yielded 1,500 to 2,000 kilocalories and 45 grams of protein per day (Dirks 1978). Kiple (1984, 76–88) has calculated the nutritional content of slave diets based on the most "optimistic" scenario, that is, assuming that slave owners followed the law, which required providing sufficient nutrition to the slave population. He concluded that even then slaves suffered from protein deficiencies, and that the high salt content of dried meat and fish led to a depletion of potassium levels. The slave diet had insufficient quantities of fat, vitamin A, vitamin B1 (thiamine), B2 (riboflavin), vitamin C, and calcium.

In addition, food was not distributed evenly within a family, and adult men received larger portions of protein-rich foods than did women and children (Kiple 1984, 81). This male-biased custom contributed to the low protein and iron status of children and especially of pregnant and lactating women. Other authors have argued that the diets were in fact of lower quality than Kiple's estimates, as Kiple assumed that each slave received three pounds (about 1.5 kilograms) of meat or fish per week. But, for example, by the end of the seventeenth-century Barbados imported about 100 pounds of fish per slave per year, which meant there was only enough for each slave to receive one quarter of a pound (about 0.1 kilograms) of fish per day, or about 0.7 kilograms per week (Bean 1975).

Although they were more nutritionally deprived as adults, the Caribbean slaves probably had a better nutritional status in very early childhood than U.S. slaves. Caribbean slave mothers usually practiced prolonged breastfeeding, following original traditions brought from their home countries in West Africa. The differences in breastfeeding practices between slaves living in the United States and the Caribbean colonies resulted most likely from the different ways in which slaves were imported from Africa into these two regions. In the United States, significant imports of indigenous Africans had ended almost sixty years before the abolition of the Atlantic slave trade in 1807, which made African-born slaves the minority by the early eighteenth century (Eltis 1982). By contrast, imports of African slaves to the Caribbean lasted for a longer period, with Caribbean slave populations regularly being replenished by Africans until 1807 (Steckel 1986b). Therefore, traditional African lifestyle and cultural customs, including breastfeeding practices, persisted for a longer time in the Caribbean than the United States.

Breastfeeding among U.S. slaves lasted for about one year (Morgan 2008), but for at least two (Handler and Corruccini 1986) or even three (Kiple 1984, 45) or four (Morgan 2008) years after birth in the Caribbean colonies, and supplementary foods were not introduced before the child was one year old (Kiple 1984, 129–130). Once such foods were introduced, they were of equally bad nutritional quality as food of the U.S. slave children, but in the Caribbean a child was still nursed, and its poor diet was improved by some amount of breast milk. In later childhood and adolescence, however, Caribbean slave children had much less rapid catchup growth than the U.S. slave children (Steckel 1994), suggesting an even more dramatic decline in nutrition after weaning.

Age at menarche can be used as an indicator of the nutritional status of female children. Data on the menarcheal age of slave girls do not exist, but James Trussell and Richard Steckel (1978) have estimated from growth spurt data that the age at menarche in the U.S. slaves did not occur later than fifteen years of age. Menarche in the Caribbean slave girls has been estimated to have occurred as much as two years later (Higman 1979).

If indeed fertility in Caribbean slave women was low, it might have resulted from long breastfeeding combined with poor nutritional status. Breastfeeding by itself has a limited ability to prevent pregnancy because it does not suppress ovarian function efficiently in women who are well fed. By contrast, long breastfeeding together with a poor diet and intense physical labor can significantly postpone resumption of ovarian function after childbirth in women with poor nutritional status. Energetic factors also negatively influence an ability to conceive in women who are not breastfeeding. In slave women, poor nutrition, high workloads, and the

119

energetic costs associated with having parasitic infections lowered the chance of conception. If an energetic burden is extreme, it can lead to complete amenorrhea. Physicians and planters frequently observed "obstructions of the menstrua" in Caribbean slave women (Kiple 1984, 110).

Work and Disease in the Caribbean Slaves

Nutritional status is not only affected by diet, but also by energy expenditure. It is generally agreed that slaves working on sugar plantations had far more demanding work regimes than slaves working in any other type of agriculture or those working in towns (Higman 1984, 179–188; Follett 2003). Some sugar plantations operated in the United States, mostly in nineteenth-century Louisiana, but they were a great deal more common in the Caribbean. Intense labor on sugar plantations led to poor nutritional status due to very high energy expenditure. Slaves simply lacked time and energy to work in their own gardens, with resulting lower caloric and nutritional intake, and further deteriorating physiological condition. Such intense labor may explain the observed differences in height between slaves from sugar and "marginal" colonies in the Caribbean.

Nutritional deprivation may have led to several diseases. Diseases are hard to diagnose from historical records, but there is some evidence that Caribbean slaves suffered from diet-related conditions. Scurvy from vitamin C deficiency, anemia from lack of iron, pellagra from lack of niacin and riboflavin, and beriberi from thiamine deficiency frequently occurred among slaves (Kiple 1984, 89–103). Infectious diseases, especially internal parasites and malaria, also can affect the rate of growth and were common in the Caribbean, especially in Guyana, Trinidad, and Jamaica (Higman 1979).

Mortality rates in the Caribbean slave population were high and show variation related to the type of crop grown in the colony (Higman 1984, 324–329). Sugar-growing colonies had 40 deaths per 1,000 people, while cotton-growing colonies had 25 deaths per 1,000. The predominant registered cause of death was diarrheal disease (mainly dysentery). Diarrheal diseases were followed in frequency by dropsy, fevers (malaria and yellow fever), tuberculosis, nervous system disease, and digestive system disease, all of which accounted for about 70 percent of classifiable diseases in the Caribbean slave population (Higman 1984, 340). Diseases were much less prevalent in the environment where U.S. slaves lived (Robertson 1996, 27).

Comparative studies suggest that the diet and disease burdens did not differ dramatically for many Caribbean slaves from the conditions they had experienced before slavery, but what changed most was the magnitude of the burden from physical labor (Kiple 1984). Slaves worked much harder than the free population in West Africa. Such increased energy expenditures greatly increased their nutritional requirements. High energy expenditure causes ovarian suppression, even in the absence of negative energy balance; however, it is likely that, at least during the seasons of the year when work was most intense, many slave women were in a state of negative energy balance. We may conclude that the biological influences of diet and physical activity on ovarian function are by themselves sufficient to explain the low fertility of Caribbean slaves.

If nutritional conditions due to low caloric intake, high-intensity labor, especially on sugar plantations, and a high incidence of infectious diseases were drastically poor for Caribbean slaves, one would expect low birth weight in black Caribbean infants to persist to the present day. As discussed previously, birth weight seems to be influenced by the nutritional conditions of many past generations. If the nutritional status of Caribbean slaves was significantly worse than among U.S. slaves, the birth weights of black Caribbean infants today should be even lower than found in black U.S. infants. The observed birth weights in contemporary Caribbean countries seem to support this hypothesis. In Barbados during the late 1950s and 1960s, the average birth weight was almost 200 grams lower than the average birth weight of infants of African descent in North America (Wells 1963). More recent data from Jamaica (between 1985 and 1989) show that the mean birth weight is 3,232 grams. During that same time period, the birth weight for European Americans was almost 8 percent (or by 270 grams) higher (Peabody, Gertler, and Leibowitz 1998). Another study from Jamaica reported the mean birth weight for newborns in 1990 as being 3,191 grams on average (Thame et al. 1997). The authors also documented that maternal stature was an important predictor of a baby's birth weight, birth length, and head size, and the size of the placenta. This result provides a strong indication that poor birth outcomes in Jamaica in the late twentieth century resulted not only from the mother's condition during pregnancy but also from the longer, multigenerational exposure to nutritional deprivation.

Birth Weight in Contemporary Migrant Populations

Is it possible that differences in mean birth weights between European Americans and African Americans have nothing do with the vagaries in history but are mostly genetic? After all, there are some genetic differences between these two populations, such as, for example, in skin pigmentation (Jablonski and Chaplin 2010). Long-term selective pressure from environmental conditions causes changes in allele frequencies and makes some traits more or less frequent than others. Is it likely that lower birth weight was selectively advantageous in the West African environment and that slaves originating from these populations had genes for "low birth weight"? If so, the same alleles will still be present in their descendants.

Most African Americans are of West African ancestry. It is estimated that about three quarters of the genetic makeup of African Americans is West African in origin, with the rest from European populations (Reed 1969; Adams and Ward 1973; David and Collins 1997). If the racial difference in birth weight reflects genetic differentiation (most likely polygenic in character), we should expect women born in West Africa to have babies with equally low birth weights as babies born to African American women. We could even predict, following the genetic difference argument, that women born in West Africa will have babies with an even lower birth weight. After all, African American women would be likely to have some of the high birth weight European genes that are not present in the West African women. In these kinds of comparisons, maternal condition during pregnancy may constitute a confounding factor, so one would wish to find populations in which the quality of maternal life would be similar. The ideal setting to test this hypothesis would be provided by comparing only women currently living in the United States. One study has done precisely that.

Infants of U.S.-born white women, Africa-born black women, and U.S.-born black women have been compared (David and Collins 1997). Of course, these populations differ in many ways. White women, on average, have a higher socioeconomic status, but this factor can be taken into account and controlled statistically. Africa-born black women also have intergenerational experience of poor conditions, but, as discussed earlier, these conditions were not as bad as the conditions experienced by the slave populations in the United States. In many African countries, nutrition is poor and the prevalence of diseases high, but physical work is much less intense, and the levels of psychological stress are much lower. On average, white women had babies weighing 3,446 grams at birth, and U.S.-born black women had babies weighing 3,089 grams. The Africa-born women had much heavier babies than the U.S.-born black women with an average birth weight of 3,333 grams. Moreover, the overall distributions of birth weights were almost identical for white and Africa-born black women, meaning that these two groups had the same percentage of births occurring in each category of birth weight (for example, women in both groups had about 15 percent of babies weighing 3,000 grams). The distributions of birth weight among U.S.-born black women was different—the whole curve was shifted to the left, indicating that these women had a higher probability of giving birth to low-birth-weight babies and a lower probability of having babies with higher weight than two other groups of women.

Two interesting multivariate analyses were performed in this study. The first showed that, after statistically controlling for differences between these two groups in age, education, marital status, number of pregnancies, prenatal care, and history of fetal loss, white women's babies still weighed 98 grams more than the babies of Africa-born women. A similar analysis, after controlling for the same major risk factors associated with low birth weight, compared the babies of white women to those of U.S.-born black women and showed that there was a remaining 248-gram difference in average birth weight.

The data were also reanalyzed to include only women who were at the lowest risk for having babies with low birth weight. These women were between twenty and thirty-nine years of age, began prenatal care in the first trimester, had at least twelve years of education, and were married to partners who also had at least twelve years of education. In this social category, the differences in mean birth weight between infants of U.S.-born white and Africa-born black women became even less pronounced. At the same time, the gap between the birth weights of infants of U.S.-born white women and U.S.-born black women did not change! It is also important to note that the proportion of infants with very low birth weight (less than 1,500 grams) was similar in both groups of black women. This detail is relevant because it suggests that the mean birth weight was not affected by the fact that one group was (for whatever reason: physiological, anatomical, or environmental) more inclined to give birth to small and premature babies. This study is very suggestive that genetic factors have little impact as a cause of the dramatic, persistent differentiation in birth weights between white and black women in the United States.

In a comparison study of U.S.-born black women and foreign-born black women from Africa, Cape Verde, the Caribbean, and England who gave birth in Boston City Hospital, the infants of foreign-born black women had higher birth weights (Cabral et al. 1990). The women had some lifestyle differences—for example, in the use of cigarettes, alcohol, marijuana, and cocaine during pregnancy. All these behaviors were more common among the U.S.-born women, so they could have had some negative impact on birth weight. When the analyses were repeated with these risky behaviors statistically accounted for, the foreign-born black women still had babies weighing 135 grams more than the U.S.-born black women.

Of course, when we are interested in the question of whether the low birth weights of African Americans are genetically determined, the proper and meaningful comparison is between U.S.-born black women and Africa-born black women. The 135- to 150-gram difference in their average birth weights that was still observed after taking into account the differences in their socioeconomic factors may, with full justification, be called slavery's brand mark.

The Recalcitrant Birth Weight

It is hard to predict how many generations with improved nutrition and better health are needed for a significant increase in birth weight. In the second half of the twentieth century, the mean birth weight of the white population in the United States increased. The weight of female infants born to U.S.-born white women rose from 3,309 grams for cohorts born in the years 1956 to 1975, to 3,374 grams for cohorts born between 1989 and 1991. This was an increase of 65 grams in twenty-five years, or 26 grams per decade. During the same period, the birth weight of female infants born to U.S.-born African American women was not only substantially lower but increased only by 17 grams (from 3,060 to 3,077), or less than 7 grams per decade. The 1956–1975 cohort's birth weight for female infants of Caribbean-born women who were living in the United States at the time of birth showed no statistically significant change (Collins, Wu, and David 2002).

Studies of populations migrating from countries with poor economic status to countries without nutritional shortages show that birth weight is rather resistant to increase. In England over the last forty years, there has been no secular trend in birth weight for babies born to mothers from the Indian subcontinent. The mean birth weight of infants born to mothers who were themselves born on the Indian subcontinent but conceived after having immigrated to England was virtually identical to the birth weight of babies of mothers from the same ethnic group who had been born in England (3,120 and 3,119 grams, respectively—only singleton births delivered at term were analyzed in this study; Margetts et al. 2002). A similar lack of difference in birth weight was documented by another study from England that compared first and second generations of Asian women (3,044 grams and 3,022 grams, respectively; Draper, Abrams, and Clarke 1995).

Other ethnic groups also showed no statistically significant change in birth weight after immigrating to the United Kingdom. Black Caribbean mothers who were born in the Caribbean had female infants weighing 3,129 grams and male infants weighing 3,320 grams compared with the children of black Caribbean mothers born in the United Kingdom, who weighed 3,223 grams and 3,275 grams at birth (female and male, respectively). A similar lack of improvement was noted for black African mothers in the birth weights of their children.

Within the United States, a study based on data from the 1979 National Longitudinal Survey of Youth and the 1970 U.S. Census showed that for European American women who were poor as children socioeconomic upward mobility contributed to improved birth weights. In contrast, similar positive changes in socioeconomic status occurred among African American women but had no effect on the proportion of infants who were born with low birth weights (Colen et al. 2006).

There are, of course, differences in the mean birth weight among babies born in countries with differing overall levels of economic development. For example, South Asian babies born in the United Kingdom are significantly heavier, by about 300 grams, than South Asian babies born on the Indian subcontinent (Rao et al. 2001). The environment of Western Europe is characterized by a lower load of infectious diseases and improved maternal nutrition, and this overall improvement in the quality of life may explain the observed differences in birth weight (Margetts et al. 2002). However, what is interesting is that, as was discussed earlier, the initial, positive response in birth weight is not followed by further increases in the ensuing generations: babies of migrant women remain significantly lighter than babies of the U.K.-born white women. All these observations clearly show that the intergenerational component of birth weight is very resistant to current improvements in the quality of nutrition. The attenuation of the signal from past generations is slow.

Is Low Birth Weight of African American Babies a Mark of Slavery?

Currently, the birth weight of African Americans is significantly lower than the birth weight of European Americans. When variation in socioeconomic factors and differences in exposure to risk factors are accounted for, a significant racial difference in birth weight still remains. Furthermore, this difference is unlikely to be genetic in origin because contemporary black women who live in the United States but were born in the African countries ancestral to slave populations give birth to children with much higher weights than black women in the United States who are descended from slaves (Cabral et al. 1990; David and Collins 1997).

Birth weight is determined by many factors, including influence of intergenerational life conditions, especially for the female line. Slavery was associated with poor nutritional status in utero, childhood, and adulthood. Poor nutritional status resulted from an inadequate diet and excessive work, an imbalance between energy intake and expenditure, and the energetic costs of fighting infectious diseases. I have suggested elsewhere (Jasienska 2009) that the low birth weight of contemporary African American children has at least three major causes, all related to the years of slavery from the past (Figure 5.1).

First, slave children had low birth weights as a direct consequence of their mothers' undernutrition and intense physical labor. Mothers experienced these conditions both during prepregnancy and during gestation. Second, their mothers had been underfed and had had to endure heavy workloads since they were young. This negatively influenced their rate of childhood growth, and a mother's slow childhood growth is one of the determinants of low birth weight in her children. Third, as birth weight is influenced by maternal (and grand- and great-grandmaternal) birth weights, slave mothers had low-birth-weight children as a result of their slave mothers, grandmothers, and great-grandmothers having had poor nutritional status during their development and adult life, including during their pregnancies.

All these factors may have had combined impact in producing a powerful, interactive joint effect on birth weight. For example, a girl born

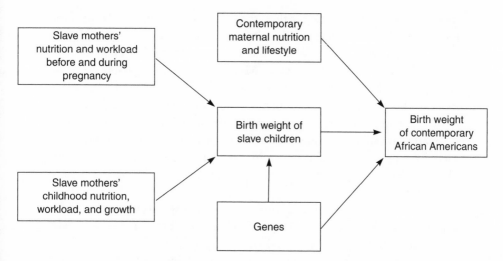

Figure 5.1. Poor maternal nutritional status and poor children's nutritional status during the years of slavery are important intergenerational factors in the low birth weights of contemporary African American babies.

with low weight will most likely grow slowly in childhood. Children with low birth weight are more prone to diseases early in life, and a sick child often receives poor nutrition as a result of lack of appetite or inability to seek additional food. Therefore, a child will have a low birth weight as a result of both her mother's low birth weight and poor growth during her own childhood.

Even short-term nutritional deprivation in pregnant women, when very severe, may reverberate across more than one generation, producing an intergenerational effect. Dutch babies exposed to famine as fetuses in mid- and late gestation had reduced birth weights. Remarkably, the effect was also detectable years later: when they became adult women, they had children whose birth weights were also reduced. Firstborn children from the second generation whose mothers had been undernourished as fetuses in the last trimester of pregnancy (i.e., they were the grandchildren of women who actually suffered through the World War II famine) still had lower birth weights (Stein and Lumey 2000), despite the fact that their mothers did not experience any malnutrition—they had adequate diets and good nutritional status while living in the modern Netherlands.

African Americans suffered nutritional deprivation that lasted much longer, both within each generation and across generations. Even though their caloric intake was higher than that of women during the Dutch famine, slaves' levels of energy expenditure were extreme. Hard work alone during pregnancy is capable of reducing an infant's birth weight, regardless of the mother's caloric intake. Multigenerational exposure to harsh energy-related conditions may change the maternal physiology's assessment of the quality of environmental conditions. Even when the mother is well nourished herself, as an organism she receives an additional intergenerational signal. The signal may be integrated into her own maternal metabolic processes, and it may cause her organism to follow a specific physiological strategy. This strategy results in the reduced birth weight of her children.

The exact physiological mechanisms are unknown at present, but there are two possible but not mutually exclusive evolutionary reasons for why a nutritionally deprived maternal organism would follow a strategy of reduced allocation of nutrients and energy to a fetus, especially under the influence of a signal of poor environmental quality in past generations. First, it may be advantageous to a mother to have a child—and consequently an adult offspring—with a reduced body size. A small body size is energetically less expensive to maintain, and thus has improved survival under conditions of low energy availability. Second, maternal fitness may be best served by investing less in a given offspring in expectation of worsening energetic conditions in the future. Based on a multigenerational signal, a woman's body "expects" that, in general, nutritional conditions will continue to be consistently poor. By investing less in the current reproductive event, the mother saves energy for future reproduction.

It is hard to determine whether producing a child of small size at birth is an adaptively advantageous, functional, strategic choice or a strategy that merely makes the "best of a bad job," as the best the mother can do in a given situation. There are obvious trade-offs associated with being small at birth. A smaller body size might allow an individual to survive on limited resources, but the mortality of children who are small at birth is very high, especially early in life. Such infants succumb more easily to infections, and they sometimes have developmental problems that negatively affect their future development. Even in the contemporary United States, small size at birth is still associated with a higher probability of death. In the conditions under which slaves were forced to live, having a child with a small size at birth was an extremely risky maternal strategy, and likely to result in a wasted energetic investment.

The hypothesis that I present in this chapter is that too few generations have elapsed for African Americans living in improved energetic status to counteract the tragic multigenerational effects of nutritional deprivation. Can we think of an alternative explanation for low birth weights among African Americans? American studies investigating variation in birth weight generally carefully control for socioeconomic and lifestyle differences between racial groups, but perhaps there are other factors that are potentially relevant but more difficult to assess quantitatively.

Some researchers have suggested that the psychological effects of racial discrimination still reported by many African American women in the contemporary United States may be responsible for the low birth weights of African American children (Rich-Edwards et al. 2001; Collins et al. 2004; Mustillo et al. 2004). Self-reported discrimination has recently achieved attention as yet another factor that may have an impact on health; for example, it may have a negative influence on mental health and health-related behaviors (Paradies 2006). At present, convincing evidence supporting a relationship between racial discrimination and low birth weights is missing, but a few studies have suggested a potential link. Cases of self-reported racial discrimination were found by a small study of poor African American women to be associated with giving birth to extremely low birth weight infants below 1,500 grams (Collins et al. 2004). Higher risk of preterm delivery in women who complained about racial discrimination was also reported in some (Dole et al. 2004; Mustillo et al. 2004) but not all (Rosenberg et al. 2002) studies.

Studies on self-reported racial discrimination have often been criticized for serious methodological limitations, including a lack of clear definitions of racism and for confounding racism with psychological stress (Paradies 2006). But when they improve their methodology, future studies may help resolve the mystery of low birth weight in African Americans.

Chapters 3–5 illustrated how an individual's health depends on the energetic conditions (i.e., nutrition, workload, disease) experienced not only by that person, not only by that person's mother, but also by recent ancestral generations. Birth weight and health in childhood and adult life are strongly affected by energetic conditions. Now we will switch our attention to how reproduction affects not the child but the child's mother.

Birth size and the biological quality of a baby depend strongly on the mother's condition because reproduction is costly. Growing the fetus, producing milk for the infant, and providing food for the older child all require a great deal of maternal energy. When energy is limited, it is not only the child's condition that is compromised—women also pay the price of reproduction. In Chapters 6 and 7, I will discuss the costs of reproduction, the strategies that women use to meet those costs, and the long-term consequences of reproduction for maternal health and longevity.

6

THE PRICE OF REPRODUCTION

In poor environments with inadequate diets, intense workloads, and a high burden of disease, it is not easy for individuals to maintain energy balance and health. Reproduction adds additional costs and makes the task of surviving even more challenging. The costs of reproduction are higher for women than for men due to the energetic and nutritional requirements of pregnancy and lactation, and to traditionally female-oriented childcare.

In New South Wales (Australia) between 1898 and 1902, women who had six children lived longer than those with larger families. The investigator explained these results, published in 1905, as a result of "incessant strain upon the physique of women who bear large families during the periods of gestation, parturition and lactation" (Powys 1905, 244).

When energy and nutrients are used to support reproduction, less is left for other purposes. This is the main principle of life-history theory. Energy and nutrients from dietary intake are used to support physiological maintenance (i.e., all physiological processes, including immune function, renewal of tissues, neutralization of toxins, and digestion) and physical activity. Growth processes in children and adolescents require substantial amounts of energy and nutrients. In adult females, pregnancy and lactation render additional costs.

Competition for resources results in trade-offs among life history traits. When resources are available in limited quantities, allocation to one function can only occur at the expense of other functions (Zera and Harshman 2001). Because reproduction is costly but from an evolutionary point of view, of course, essential, the costs of reproductive effort must be paid. Increased allocation to current reproduction necessarily results in reduced future reproduction and/or reduced life span. This logic is not questioned, but the mechanisms generating such trade-offs are not well understood.

Pregnancy and lactation do take energy and nutrients away from other processes, but exactly what kind of physiological functions are negatively affected by reproductive effort?

Fruit flies *(Drosophila)* forced to lay more eggs than usual have a decreased resistance to in vivo exposure to free radicals (Salmon, Marx, and Harshman 2001). Similarly, the red blood cells of zebra finches, when the bird has been forced into higher reproductive costs, show reduced resistance when exposed to free radicals in vitro (Alonso-Alvarez et al. 2004). In mammals, the costs of milk production involve trade-offs: for example, lactating bighorn sheep *(Ovis canadensis)* have an increased parasite load (Festa-Bianchet 1989).

But the costs of reproduction are mainly expected in circumstances when resources are limited and inadequate to support competing processes. In a different study of zebra finches, a negative correlation between the number of laid eggs and resistance to oxidative stress was documented, but interestingly, this relationship was not present in the birds whose diet was supplemented with carotenoids (Bertrand et al. 2006).

Energetic Costs of Reproduction

Reproduction in females requires energy to maintain ovarian and uterine function (Johnson et al. 1994). Additional energy beyond regular metabolic needs is also necessary to support regular menstrual function (Strassmann 1996). However, the energetic costs of menstrual cycles are negligible in comparison to the much higher and longer lasting demands of pregnancy and lactation. The maintenance of menstrual cycles requires only a 6 percent to 12 percent increase in the resting metabolic rate, which adds about 70 to 140 kilocalories per day to regular energy requirements (Bisdee, James, and Shaw 1989; Meijer et al. 1992; Howe, Rumpler, and Seale 1993; Curtis et al. 1996). Women are known to boost their energy intake during the days after ovulation, most likely to meet this additional energetic demand (Johnson et al. 1994). However, these higher expenses occur only for a few days during the luteal phase of the cycle.

Human mothers have babies that are more costly to produce than the babies of our closest primate relatives (Haig 2008). The body weight of a woman is only a little higher than that of a chimpanzee female (Grether and Yerkes 1940), but the average birth weight of human infants is about 3.5 kilograms while chimp infants weigh only about 2.0 kilograms

(Smith, Butler, and Pace 1975). Human babies are also much fatter than the babies of most other mammalian species (Kuzawa 1998). For well-nourished women from industrialized countries, the estimated costs of pregnancy constitute a daily expense of, on average, an extra 90, 290, and 470 kilocalories per day for the first, second, and third trimesters, respectively (Butte and King 2005). During the last trimester, the woman may require up to 22 percent of additional energy over prepregnant values (Hytten and Leitch 1971; Butte et al. 1999). The total cost of pregnancy for women with a mean weight gain during gestation of 12 kilograms is approximately 77,200 kilocalories (Butte and King 2005).

The energetic costs of lactation are even higher (Dufour and Sauther 2002). They change with the age of the infant (Rashid and Ulijaszek 1999) and the frequency of feeding, but on average lactation requires an additional 626 kilocalories per day (Butte and King 2005) and may last a few years. These estimated costs have been criticized as too high in comparison with the actual energy intake of pregnant and lactating women from non-industrial populations (Illingworth et al. 1986; Kopp-Hoolihan et al. 1999). Well-nourished British women had an average energy intake of 1,980 kilocalories per day when pregnant, and over 2,270 kilocalories per day while lactating (Prentice 1984). In Swedish women, the total energy was intake was 2,870 kilocalories per day during the period of exclusive breastfeeding (i.e., when a baby drinks only mother's milk) of three-month-old babies (Butte, Wong, and Hopkinson 2001). In comparison, rural Gambian women during the wet (hunger) season supported pregnancy and lactation on an average energy intake of only 1,290 kilocalories per day (Prentice 1984).

It should be stressed that women are capable of supporting reproduction on a limited energy supply, but such a strategy entails substantial long-term costs. Reproduction in women who are in poor energetic condition is associated with a diminished reproductive outcome, reflected both in the health of the newborn child (Lechtig et al. 1975; Roberts et al. 1982; Kusin et al. 1992; Pike 2000) and in a further decline in the nutritional status of the mother, known as *maternal depletion syndrome* (Merchant and Martorell 1988; National Academy of Sciences Committee on Population 1989; Tracer 1991; Little, Leslie, and Campbell 1992; Miller, Rodriguez, and Pebley 1994; Khan, Chien, and Khan 1998; Winkvist, Habicht, and Rasmussen 1998; Pike 1999; George et al. 2000). Such depletion, by definition, negatively affects her future reproductive potential.

We discussed earlier the phenomenon of ovarian suppression that helps women avoid pregnancy when their nutritional status is poor. Some women, however, have poor nutritional status throughout their lives, so an ability to temporarily suppress ovarian function is particularly important for them. Ovarian suppression occurring during tough times lowers the probability of conception and lengthens the interval between births (Ellison 2003a). Thanks to such a mechanism, the mother has more time to recuperate after pregnancy and lactation. Improvement in her nutritional status results in higher levels of ovarian hormones, a higher probability of conception, and, consequently, a new pregnancy. In chronically poor environments, allocating energy to producing a new child will require implementing various behavioral and physiological strategies aimed at improving efficiency and lowering the costs of reproduction.

Ways of Allocating Energy to Reproduction

If energy and nutrients are limited (as was the case throughout our evolutionary past and is the case in many populations today) and reproduction magnifies demands for these resources, how are women able to do it all—maintain good health, survive, and produce children? One way of meeting the additional demands of reproduction is by increasing energy intake. Eating more, what is commonly called "eating for two" in pregnancy, is an obvious way to meet additional demands. In many populations, however, this is not possible. People have limited financial means and may not be able to increase their dietary intake. In our evolutionary past it was possible, at times, to increase dietary intake, but it probably involved raising levels of physical activity as well. Searching for more food on gathering trips meant walking longer distances, a longer time of digging for roots, and heavier loads carried back home.

Work Less, Rest More

Another way of saving energy and redirecting it to reproduction, is to work less. Some pregnant and lactating women are indeed able to decrease their physical activity. However, whether this is possible depends on the lifestyle and often the season of the year. In agricultural populations, during seasons with high labor demands, such as during the harvest, women may not be able to work less when pregnant because every pair of hands is needed to ensure that enough food is collected.

134

In rural Poland, in the small agricultural villages that for the past fifteen years have been part of the Mogielica Human Ecology Study Site, the cereal harvest occurs usually in August and lasts for two to three weeks. A single family works their own fields. The harvest is weather dependent, and cereals cannot be cut and collected when it is raining. If the weather is good and the cereals are ripe, all members of the family work, including children, grandparents, and pregnant women. Anna, now eighty years old, recalled one of her pregnancies. On a beautiful summer day, she was pregnant and very close to delivery—but so was the wheat. It was ripe and ready for harvest, so Anna worked in the fields from early morning. She had already given birth to several children, so when she felt pain, she knew she had to go back home. Her husband ran to find a local midwife and went back to the field when she arrived. Anna gave birth to the child, left the newborn at home with her older daughter, and returned to the fields to finish the harvest.

Anna's story may be an extreme example, but it shows how difficult it is for a woman to remove herself from the demands of everyday work, especially during times like harvest. Anna knew that the harvested cereal would feed her and her children for the entire year. Her family was poor, and they could not afford to hire a worker to replace Anna. Older children worked as well, but that was not enough. Children provide a lot of help in agricultural communities, but many tasks associated with agricultural work require skill, experience, and physical strength. Women from many societies face similar challenges and cannot reduce their physical activity when meeting the high energetic demands of reproduction.

A review of 122 studies showed that in developing countries, in most societies, women are expected to continue their work routine throughout most of pregnancy (Institute of Medicine 1992). In a small village in the Philippines, pregnant women are supposed to work as usual until the time of delivery (Raphael and Davis 1985, 30). Pregnant or lactating women in rural Nepal cannot reduce their workload during the spring or monsoon season when work is very demanding; their levels of energy expenditure are just as high as those of women who are not pregnant or breastfeeding (Panter-Brick 1993).

Fat, but Not Fat Enough

Humans, just like many other species, have the ability to store metabolic energy. When times are good—meaning when food is abundant and work is not very intense—people consume more energy than they expend. The surplus of metabolic energy is stored as fat. Fat stores can be used and metabolized to provide energy when not enough is coming from dietary consumption. Women are physiologically more "skillful" at storing fat than men, and many researchers believe that the reason for that ability is related to the high costs of reproduction. Can women use the stored energy when they need more than they can obtain from food? Of course they can.

People of both genders burn fat when their energy expenditure is higher than their energy intake. Such an energetic state is called a *negative energy balance*. In women, fat is stored during pregnancy in preparation for lactation. It has been suggested that this fat storage system, which is unique to humans (most primates support lactation via increased food acquisition during the lactation period), evolved to provide an uninterrupted and nonfluctuating energy flow to the fast-growing infant's brain (Lancaster 1986; Kaplan 1997).

Fat stores may indeed be crucial for partially covering the energetic costs of milk production (McNamara 1995), but the important question is whether the energy stores are sufficient to support the entire costs of reproduction. Although stored energy is very helpful in meeting these costs, there are reasons why it is not sufficient to cover the entire costs of pregnancy and lactation.

First, we need to remember that well-fed (and often overfed) women from Western countries do not represent the physiological norm. As discussed previously, when we analyze evolutionary adaptations it is important to think about our ancestors. All adaptations humans have are the result of natural selection operating on thousands of the past generations. Women from our evolutionary past, our hunting-gathering great-grandmothers, clearly did not have large energy stores. The Paleolithic hunting and gathering lifestyle did not allow the luxury of substantial fat storage. Their diet, while most of the time sufficient in calories, hardly ever had caloric excess, and their levels of energy expenditure were relatively high. Modern hunter-gatherers are almost never fat; only those in a few populations, especially in South America, have visible fat accumulation.

Second, the data showing that women in developing countries often have very low fat reserves (Lawrence, Lawrence et al. 1987; Little, Leslie, and Campbell 1992; Panter-Brick 1996) have led to the hypothesis that in the developing world the function of fat stores is to serve as an emergency resource for times when conditions become very hard, rather than to steadily support milk production (Prentice and Prentice 1990; Lunn 1994). Studies of women in the rural Gambia provide support for this thinking. If fat reserves were to be used by Gambian women to subsidize just 50 percent of the costs of lactation, they would last for only four months (Prentice and Prentice 1990). In contrast, fat stores would last for eleven months when used to support half of lactation costs in well-nourished Western women. When lactation lasts for a long time, fat reserves, especially in women of poor nutritional status, are vastly insufficient to cover the costs of milk production.

Finally, in an unpredictable environment, a mother cannot use too much of her stored energy because she does not know when she will have an opportunity to replenish her energy storage. She cannot invest too much in her current offspring because that may negatively affect her future reproductive career.

Stealing Energy from Herself

The last potential source from which energy can be allocated to reproductive processes is physiological maintenance. The daily costs of physiological maintenance in humans are high. They are expressed in what bioenergetics refers to as the basal metabolic rate (BMR). The average BMR for women from several foraging, pastoral, and agricultural populations is about 1,260 kilocalories per day (Leonard 2008). Can women decrease their daily metabolic requirements and save some energy for reproductive processes?

The costs of pregnancy result from fetal growth, growth and maintenance of maternal supporting tissues, and accumulation of maternal fat (Hytten and Chamberlain 1991; Blackburn and Loper 1992). The energetic demands of lactation are a consequence of milk synthesis and the maintenance of metabolically active mammary glands (Trayhurn 1989; Prentice and Prentice 1990; Lunn 1994). Because the BMR is the index integrating all metabolic activities, it is expected to rise in women who are pregnant or nursing a child. The BMR of pregnant woman should increase during each of the four quarters of pregnancy by about 3, 7, 11, and 17 percent, respectively, above prepregnant values (Hytten and Leitch 1971). Empirical data usually support these theoretical predictions but only in women who have a good energetic status (Prentice and Whitehead 1987; Prentice et al. 1989; Prentice and Prentice 1990; Durnin 1991, 1993; Prentice et al. 1996; Butte et al. 1999). In Swedish women, the BMR increases from the very beginning of the pregnancy (Prentice and Whitehead 1987), and the cumulative increase in BMR for the entire pregnancy is over 50,000 kilocalories (210 megajoules) (Butte and King 2005). In well-nourished, healthy women the average increases in BMR were 4.5, 10.8, and 24.0 percent over pregnant values for the first, second, and third trimesters, respectively (Butte and King 2005).

However, raising their BMR when pregnant is not an option available to all women. When environmental conditions do not provide adequate energy to support the energetic demands of the child, the physiology of women who are already pregnant or lactating may involve switching on various energy-saving "procedures" (Peacock 1991). One mechanism that frees some energy to be reallocated to reproduction is a reduction in maternal BMR (Prentice et al. 1989; Peacock 1991; Prentice et al. 1995).

Reduction in BMR in pregnant or lactating women is indeed observed. A decrease in BMR for up to the twelfth week of pregnancy occurred in women from Scotland and the Gambia who were in poor nutritional condition. After the initial decline, the BMR of these women approached prepregnant values as late as in the twenty-second and twenty-sixth weeks, and started to rise further; however, even at delivery it was still much lower than the BMR of well-nourished Swedish women (Prentice and Whitehead 1987). The cumulative change in BMR for the entire pregnancy for Gambian women showed an almost 11,000 kilocalorie (45 megajoule) *decrease* in comparison with their BMR from before pregnancy (Butte and King 2005). Pregnant Nigerian women also showed significant variability in their basal metabolic rates, which corresponded to variability in their nutritional status (Cole, Ibeziako, and Bamgboye 1989).

Even well-nourished Western women show significant variation in the responses of their basal metabolism to pregnancy. A higher increase in BMR occurs in women who were in better nutritional condition before the beginning of pregnancy, as indicated by their prepregnant body fat (Prentice et al. 1989). During pregnancy, maternal adipose reserves serve as a highly significant predictor of changes in BMR (Bronstein, Mak, and King 1996). In other words, an increase in BMR occurs in women who can energetically afford it.

A reduction in BMR at the beginning of pregnancy is therefore of crucial importance for women who are in poor nutritional condition. Lowering BMR not only helps to support pregnancy by considerably reducing energy intake, but it also allows women to reallocate some energy into fat storage. This energy may later be invaluable in helping to cover the energy costs of lactation.

Lactation is predicted to cause, on average, a 12 percent increase in BMR above nonpregnant values (Hytten and Leitch 1971). Just as in the case of pregnancy, BMR does not always increase according to theoretical predictions. During lactation, BMR has been observed to increase, decrease, or remain at prepregnant levels (Lawrence and Whitehead 1988; Goldberg et al. 1991; Forsum et al. 1992; Guillermo-Tuazon et al. 1992; Madhavapeddi and Rao 1992; Piers et al. 1995). Variation in BMR response also here can be explained by differences in the nutritional status of the women studied (Prentice and Prentice 1990). During the first year of lactation, women from the Gambia showed a 5 percent decrease in BMR compared with their prepregnant values (Lawrence and Whitehead 1988). Although this seems like a very small reduction in BMR, it saves about 120 kilocalories (500 kilojoules) per day, and, in comparison with women who increased their BMR during lactation, the savings add to about 215 kilocalories (900 kilojoules) per day. The saved energy can be allocated into synthesis of milk, thus helping the mother who has low energy intake to support her child. The savings from the reduction of BMR, although not sufficient to cover the whole cost of breastfeeding, are still significant because the average cost of milk production for most women from poor populations is about 480 kilocalories (2,000 kilojoules) per day (Lunn 1994).

A reduction in BMR indicates that some metabolic processes (e.g., some components of maintenance metabolism such as protein turnover, fighting free radicals, or maintenance of immune function) are temporarily slowed or even halted (Prentice and Whitehead 1987; King et al. 1994). Not surprisingly, a long-term reduction in BMR is not cost-free and may even be detrimental to maternal health.

In general, reproduction in women who are in poor nutritional status is characterized by marked decreases in BMR during pregnancy and lactation. The BMR reduction is important as an energy-saving mechanism, especially in women who are not able to sufficiently increase their food intake while pregnant or lactating. According to the predictions of life-history theory, such trade-offs between maternal BMR and the energetic requirements of the offspring must be associated with a cost to the maternal organism.

More on Energetic Costs: Comparing
Three Reproductive Histories

The first wife of Fyodor Vassilyev (1707–1782) of Shuya, Russia, to date holds the record for the greatest number of children allegedly ever born to one woman (*Guinness Book of World Records* 1998). In forty years, she was reputed to have had twenty-seven pregnancies from which the amazing number of sixty-nine children were born. She gave birth to sixteen pairs of twins, seven sets of triplets, and four sets of quadruplets. Nothing is known about her life, and we can only assume that she had very high lifetime energetic expenses associated with her pregnancies.

Basic calculations of the costs of reproduction in individual women can be based on the total number of pregnancies and on the costs of lactation. The cost of lactation is a very important variable in calculations of total reproductive costs because, on the average, one day of lactation places higher demands on maternal energetics than one day of pregnancy. These costs are difficult to obtain, however.

It is, in principle, possible to collect information about the total duration of lactation and the frequency of daily nursing episodes; it is unlikely that mothers remember this with any precision, especially if they have had more than a few children. In addition, when a mother begins to supplement her child's diet with other food, the supplemented child will not need as much milk and will drink less during each breastfeeding session. Lower milk consumption means lower milk production and lower energetic costs for a woman. So, although information on lactation is not present in historical records, for many populations and time periods the duration of lactation can be estimated based on general knowledge about the lifestyle of that population and the availability of supplementary foods.

In addition to the direct costs of pregnancy and lactation, there are also significant indirect costs associated with childcare or with an increase in the intensity and duration of work to obtain the resources necessary to support the family. These costs are not easy to calculate either. Some anthropologists have pointed out, for example, that carrying an infant is a heavy energetic burden for a mother (Wall-Scheffler, Geiger, and Steudel-Numbers 2007). Women who do not have any additional help clearly spend more time and energy providing for each additional child. However, as children grow older, they can provide childcare to younger siblings as well as participate in housework, agricultural work, or work for

140

wages. Therefore, the relationship between the number of children and the maternal costs of childcare is clearly not linear.

Even if the total costs of reproduction based on the costs of pregnancy, lactation, and childcare are calculated, this is just part of the story when we attempt to assess the lifetime impact of costs of reproduction on maternal health and longevity. When female zebra finches are forced to pay the high costs of reproduction by laying more eggs, they do not seem to suffer negative metabolic consequences, provided that their diets are supplemented. It is likely that the same is true for human females. High costs of reproduction will not have the same effect on women who have good diets and can afford low levels of physical activity.

Emily Lennox was born in 1731 to a British aristocratic family (Tillyard 1994, 13). She had her first child when she was sixteen years old and the last when she was forty-seven. During these thirty years of reproductive life, Emily gave birth to twenty-two children. She had several more pregnancies but miscarried in her mid-forties. Contraception was not used, and sexual abstinence was the only method to avoid pregnancy. During 1777, when forty-six-year-old Emily gave birth to a baby daughter and had three miscarriages, her second husband "insisted that they part for several months, saying that they could not risk another pregnancy and that he could resist passion's call only if they put at least one sea between them" (Tillyard 1994, 290).

Such high fertility should lead to high lifetime reproductive costs and have a negative effect on health and longevity. However, Emily, like most other aristocratic women of her time, never breastfed any of her children (Tillyard 1994, 207). Her family was very well off, and she and her first husband employed almost 100 servants and staff (Tillyard 1994, 169). Thus, she could afford a diet high in calories and nutrients. The most energy-requiring component of her physical activity was horseback riding, which, to her regret, she had to give up when pregnant. Wet nurses breastfed her infants, nannies took care of the young children, and governors tended to the older ones.

In contrast, Anna was born to a poor family in a small village in southern Poland shortly after the World War I. Marriage did not improve her economic status: she and her husband inherited about twelve acres of agricultural land, but their fields were scattered on the slopes of two mountains and had poor soil, so their cultivation required strenuous manual labor. Their house had both stables for animals and living quarters for people under one roof. Until the late 1970s when they built their new house, all they had was a large, single room that served as kitchen, bedroom, and living room.

Their diet was simple; during the spring, when food from the last harvest was almost gone and the newly planted food was not yet ready to eat, their diet was deficient in calories and nutrients. The agricultural work was seasonal, with very intense labor during the summers, but the housework, animal care, and childcare were also energy demanding and did not vary much throughout the year. Even going to church on Sundays required walking two kilometers each way. Anna gave birth to twelve children; one died soon after birth, and another in later childhood. Eleven of her children were breastfed for about a year each.

We can contrast the lives of Emily Lennox and Anna with that of Nisa, a member of the hunting-gathering population of !Kung Bushmen living in the Kalahari Desert in Botswana, whose life story was beautifully described by the late anthropologist Marjorie Shostak (1983). It is interesting to quantitatively compare the reproductive energetic costs for these three women—an aristocrat, a rural farmer, and a hunter-gatherer—even though such calculations cannot be very precise (Table 6.1).

There are many things we do not know. When it comes to pregnancy, only those that ended in childbirth are usually recorded. The durations of the pregnancies are not known; it is likely that there was some variability in pregnancy length. On the other hand, variability in pregnancy lengths certainly was less pronounced than the variability present in modern, industrial societies. Without medical technology, preterm babies had a low chance of survival, so only full-term or almost full-term gestations resulted in live births, with infants surviving longer than a few weeks. Therefore, all the calculations used here for pregnancies are for full term.

The costs of lactation are even more difficult to calculate with precision. We can assume an average duration of lactation of twelve months per each child for rural farmer women, and thirty-six months per child for hunter-gatherer women (Eaton and Eaton III 1999). Of course, there is variation in length of breastfeeding within agricultural and within hunting-gathering populations. The timing of weaning may depend in rural communities on the season of the year, and a one-year-old child, who would be weaned when reaching this age in the summer, may still be breastfed on reaching this age during early spring when no weaning food is available. Hunter-gatherers, even those who have abundant food, may not have suitable weaning food in some seasons (Clayton, Sealy, and

Table 6.1. Estimated costs of reproduction, energy expenditure, and energy intake for women representing populations with different lifestyles.

	Emily Lennox Aristocrat, Ireland, 18th century	Anna Rural agriculturalist, Poland, 20th century	Nisa Hunter-gatherer, Africa, 20th century
Number of pregnancies[a]	22	12	4
Total energetic cost of pregnancies (kcal)[b]	1,697,000	926,000	309,000
Average duration of lactation (months)	0	12	36
Total energetic cost of lactation (kcal)[c]	0	2,479,000[d]	2,704,000
Total lifetime cost of reproduction (pregnancy and lactation, kcal)	1,697,000	3,405,000	3,013,000
Daily energy expenditure in nonpregnant, nonlactating state (kcal)	1,835[e]	2,558[f]	1,771[g]
Daily energy intake (kcal)	Sufficient intake	2,870[f]	2,140[h]

Note: The values of total costs were rounded to the nearest 1,000 kilocalories (kcal); the values of daily costs are given as estimated by the researchers.

a. Only full-term pregnancies are included.

b. Total cost of full-term pregnancy was estimated as 323 megajoules (77,147 kcal) (Butte and King 2005).

c. Daily cost of lactation was estimated as 2.62 megajoules/day (626 kcal) (Butte and King 2005).

d. Cost of lactation calculated for eleven children, as one died soon after birth.

e. Cordain et al. 1998.

f. Jasienska and Ellison 2004.

g. Leonard and Robertson 1992.

h. This value was given for a !Kung San adult, not necessarily a woman (Lee 1968).

Pfeiffer 2006). In the Kalahari desert, larvae and caterpillars are used as protein-rich weaning food, but they are only seasonally available. Among the nomadic !Kung, interbirth intervals are on average forty-four months long (Konner and Worthman 1980), which suggests a long period of breast-feeding. !Kung women believe that a child needs breast milk until it is at least three or four years old.

It is also likely that there are differences in the costs of lactation depending on the child's sex. Milk composition (Hinde 2009; Powe, Knott, and Conklin-Brittain 2010), enlargement of maternal breast size

(Galbarczyk 2011), and the duration of breastfeeding differ between mothers of male and female children. For example, in societies where people prefer sons, male infants are nursed for a longer time than female infants (Graham, Larsen, and Xu 1998).

The comparison of energetic costs of lifetime pregnancies and lactations for the three women introduced here, representing three very different societies, shows that the costs were highest for Anna, even though Emily Lennox had much greater fertility (Table 6.1). Anna's costs were twice as high as Emily's because Emily did not breastfeed her children. Somewhat surprisingly, the energetic costs of reproduction for Nisa, who only had four children, were almost twice as high as the costs for Emily with her twenty-two births.

The energetic costs of reproduction should be compared against the energetic costs of living. Women differ in how much energy they devote to reproduction, and they also differ in how much energy they use for physical activity. Differences in how easily they can compensate for their energy expenditures by energy intake are also important. The difference between energy intake and energy expenditure is an indication of the nutritional status of the individual. For Anna, Emily, and Nisa, the energy costs of living were never measured, but there have been measurements or estimates for both energy expenditure and intake for other women in similar societies. Emily, most likely, had the levels of energy expenditure of a modern sedentary person. Her energy intake was clearly adequate but not excessive, assuming that the paintings of her were accurate, because she was not visibly overweight. When she was in her seventies, her husband told her in a letter that she was "the most beautiful woman of your age in the kingdom" (Tillyard 1994, 369).

The values for Anna are estimates based on values of energy expenditure and intake calculated from the interviews regarding her twenty-four-hour activity and diet, which I conducted in Anna's village during my research on seasonal changes in lifestyle and ovarian function described in Chapter 1 (Jasienska and Ellison 1998, 2004). However, these data were collected at the end of the twentieth century when people no longer experienced nutritional shortages. For Anna, estimations of her energy expenditure based on such modern values may still be accurate, but her levels of energy intake would most likely be overestimated. Nutritional shortages were a fact of life after World War II in rural Poland, and caloric shortages were especially likely to occur in the spring, before new food was ready to be harvested.

For hunter-gatherers like Nisa, energy intake is closely linked to energy expenditure. To increase her intake, Nisa had to increase her expenditure. For most of her life, her diet was energetically adequate but never calorie excessive, and some seasonal shortages were also likely to have occurred. In hunter-gatherers it is extremely difficult for women with higher energy needs when pregnant or lactating to obtain more food than what they had consumed in a nonpregnant, nonlactating state.

All values for energy expenditure and intake are for nonpregnant, non-lactating women. They are not corrected for differences in body weight. !Kung women weigh only 41 kilograms, whereas the average weight of women in Anna's village is 64 kilograms. No data allow for the estimation of body weight for Emily, and body weight is important because it determines the costs of basal metabolism. Total costs of energy expenditure include both basal metabolism and physical activity. Therefore, because the body weight of a !Kung woman is low, her costs of physical activity (presented in Table 6.1) are in fact relatively higher than the costs would have been for Emily, who had a much higher body weight and thus higher BMR.

Costs of Reproduction and Long-Term Health Risks

A maternal organism must find additional energy and nutrients for growing and feeding a fetus and later an infant. In addition, the physiological and metabolic adjustments required by the reproductive episode may cause permanent changes in the maternal organism, especially when pregnancies are numerous. Are the costs of reproduction expected only for women from more traditional populations such as Anna or Nisa, or do women from economically developed countries also experience them?

In the United States, the self-reported health status is lower for women who have had at least three pregnancies, and especially in women who have had six or more pregnancies, after adjusting for age, ethnicity, education, marital status, household income, and wealth (Kington, Lillard, and Rogowski 1997). This result is meaningful because the self-reported health status, however subjective it may sound, is a reliable predictor of mortality (Idler and Benyamini 1997).

Cardiovascular Health and Diabetes

Are the high costs of reproduction really negatively affecting some aspects of health? Not all studies agree, but most evidence strongly suggests that

parity in well-nourished women is positively related to their risk of obesity, impaired glucose tolerance, non-insulin-dependent diabetes, and cardiovascular disease.

In women participating in the longitudinal Framingham Heart Study
and the National Health and Nutrition Examination Survey, the number
of pregnancies was positively correlated with the subsequent development of cardiovascular disease (Ness et al. 1993). In British women who
had at least two children, the risk of coronary heart disease rose by 30
percent with each additional child. The risk of developing heart disease
was lower but still statistically significant after the confounding influences
of obesity and metabolic risk factors were taken into account (Lawlor et
al. 2003). A high number of pregnancies also seems to be associated with
an increased risk of stroke: in U.S. women, having gone through six or
more pregnancies increased the risk of all types of strokes by 70 percent
(Qureshi et al. 1997).

A relationship between parity and the risk of diabetes has not been confirmed by all studies, but many suggest that such a relationship is real. For
example, the risk of diabetes was 42 percent higher in Finnish women who
had five children or more compared with overall national statistics (Hinkula
et al. 2006). Among rural women from Australia, those with more than
four children had a 28 percent higher risk of diabetes than women with
three or four children, and a 35 percent higher risk than women with one
or two children, after taking into account variation in obesity (Simmons
et al. 2006). Interestingly, some studies also showed an increased risk of
diabetes in childless women (Simmons 1992). This could likely be the
reason that the studies that included childless women often failed to show
a linear relationship between parity and risk of diabetes.

In women from economically developed countries, high fertility increases the risk of being overweight or obese. Body weight is gained during
each pregnancy, and most women never lose at least some of the added
weight. In a large study of U.S. women, each birth was associated with a
permanent increase in body weight even though women gained only a
modest 0.55 kilograms per each child. Further, a higher proportion of
women with a parity of three or more were overweight compared with
women who had lower parity (Brown, Kaye, and Folsom 1992). In American women from Utah, the relationship observed between the number of
children and risk of obesity was of a dose–response type: each additional
live birth increased the risk of obesity by 11 percent. This relationship was

independent of variation in socioeconomic status or other factors known to affect the risk of obesity.

In women from developing countries, the relationship between parity and obesity tends to follow a different pattern. Repeated reproductive events cause not a gain but a reduction in body weight and body fat. The concept of maternal depletion syndrome was, in fact, proposed to describe long-term negative changes in maternal nutritional status, as opposed to short-term (and easily reversible) declines associated with a single pregnancy or subsequent breastfeeding (Winkvist, Rasmussen, and Habicht 1992).

Maternal depletion syndrome has been documented in many populations living with poor nutritional conditions. Among the African !Kung San, a higher number of surviving children is associated with a lower body weight in women (Kirchengast 2000). In men, the opposite pattern was discovered: those with more surviving children had a higher body weight (Kirchengast 2000). This contrast indicates that in women a decline in body weight is likely to result from reproductive costs, and social factors in !Kung San society may magnify this effect. Among the Turkana from northwest Kenya, who are lean and frequently experience dietary shortages, women from both nomadic and settled populations had declines in fat reserves that were correlated with parity (Little, Leslie, and Campbell 1992).

In Papua New Guinea, the nutritional status of women also worsens with parity (Garner et al. 1994). Maternal depletion in this population occurs even though the birth intervals are relatively long (three years on the average). However, women gain only about 5 kilograms of weight during pregnancy, which means that their weight gain from fat is very low. Fat deposition, together with blood and extracellular volume expansion, contribute only about 650 grams of weight gain, the rest of the added weight being the weight of the placenta and the fetus itself.

Better socioeconomic status seems to improve the mother's ability to cope with the energetic and physiological burden of repeated reproductive events. In Papua New Guinea, the decline in nutritional status mentioned previously is especially pronounced in women who lead traditional lives as foraging-horticulturalists. Such negative changes have not been observed among the wage-earning women (Tracer 1991). When Rendille Kenyan women changed their lifestyle from nomadic to settled by moving to a large city, they improved their nutritional status as well.

In contrast, Rendille women who still lead a nomadic life show parity-related decreases in body mass index and fat reserves (Shell-Duncan and Yung 2004). Maternal depletion syndrome does not occur in well-nourished Toba women from Argentina, a population who are undergoing a gradual transition from seminomadic hunting-gathering to a sedentary, peri-urban lifestyle. They do not lose excess weight gained during pregnancy despite prolonged, intense breastfeeding (Valeggia and Ellison 2003).

Physiological Mechanisms in Pregnancy and Lactation Related to Risk of Diseases

Epidemiological studies in women from developed countries show a relationship between parity and risk of cardiovascular diseases, which suggests a number of mechanisms that may underlie this relationship. Early pregnancy represents an anabolic state, meaning the phase of synthesis; metabolic changes during this phase encourage lipogenesis and fat storage. Later, pregnancy, while the fetus is rapidly growing, represents the catabolic phase for the maternal body, during which time maternal fat tissue is used for energy.

Pregnancy is characterized by insulin resistance, which has an important function because it increases lipolysis in fat tissue and, as a result, enhances the flux of fatty acids to the maternal liver. This promotes the synthesis of very-low-density lipoproteins (VLDL) and an increase in triglyceride levels (Toescu et al. 2002). Although VLDL is normally removed from the body, the activity of the enzyme responsible for this removal diminishes in response to insulin resistance. This causes the accumulation of VLDL in maternal plasma and leads to an increased accumulation of low-density lipoprotein (LDL, the bad cholesterol). Lipid metabolism changes over the course of pregnancy, moving from normal or even lower cholesterol levels during early pregnancy to increased triglyceride levels in late pregnancy. Therefore, women have high levels of triglycerides and LDL cholesterol during late pregnancy.

Low-density lipoprotein is an important factor in the development of atherosclerosis. The presence of smaller, denser LDL particles promotes a higher risk, and exactly this form of cholesterol is present in a higher fraction of pregnant than nonpregnant women (Toescu et al. 2002). These smaller particles are also more susceptible to oxidation. Oxidative stress can be defined as an imbalance between free radical damage and antioxidant protection. Consequently, pregnancy can be identified as a state with increased oxidative stress, as indicated by a rise in lipid hydroper-oxides, a commonly used marker of oxidative stress (Toescu et al. 2002).

The atherosclerotic changes that occur during a single pregnancy may be relatively small, but the accumulated damage caused by repeated preg-nancies could be substantial. For example, in the Rotterdam study of over 4,800 women, parous women had a 36 percent higher incidence of athero-sclerosis than women with no children, and women with four or more children had a 64 percent higher incidence than childless women (Hum-phries et al. 2001). This relationship remained statistically significant after adjustment for lipid levels, insulin resistance, and obesity.

Insulin sensitivity decreases by about 50 percent during the last se-mester of pregnancy (Catalano et al. 1991), but it usually returns to pre-pregnancy levels during the postpartum period (Seghieri et al. 2005). It is possible, however, that repeated pregnancies may induce a progressive increase in insulin resistance that, in turn, may lead to impaired glucose tolerance and type 2 diabetes (Kritz-Silverstein, Barrett-Connor, and Wingard 1989). Data do suggest that this may be the case. In women of postreproductive age, a high number of pregnancies has been associated with increased levels of fasting insulin and decreased insulin sensitivity (Kritz-Silverstein et al. 1994).

Reproduction, Bones, and Osteoporosis

During pregnancy and lactation, calcium is required to support the de-veloping skeleton of the fetus. High calcium requirements are always met by mobilization of calcium from the skeleton of the mother (Prentice 2000). We may hypothesize that women who have had a large number of pregnancies and have breastfed their children would have lower bone mineral density, making them more likely to suffer from osteoporosis in later life.

Each pregnancy causes a 3 to 4.5 percent decrease in bone mineral density in the lumbar region (Drinkwater and Chestnut 1991; Black et al. 2000), which further decreases by 3 to 6 percent during lactation (Laskey and Prentice 1997; Karlsson, Obrant, and Karlsson 2001). These changes are substantial. For comparison with the changes that occur during reproductive events, postmenopausal women lose only 1 to 3 percent per year of bone mineral content at the spine and hip.

A decrease in density of just 10 percent is dangerous because it doubles the chance of fracturing a bone (Cummings et al. 1995). Fortunately, the changes in bone density that occur during pregnancy and lactation are reversible (Laskey and Prentice 1997), at least in women from industrialized populations. There has been no convincing evidence for a negative relationship between the number of children and bone mineral density of the mother at older ages (Bererhi et al. 1996).

Bone mineral density is positively related to the number of reproductive years, measuring the period from menarche to menopause. As estrogen has a positive effect on bones, the number of menstrual cycles during which estrogen is produced should indeed be positively related to bone mineral density (Somner et al. 2004). In fact, both androgens and estrogens are important for the development and maintenance of bone mass. Estrogen deficiency is a major culprit in the development of osteoporosis (Raisz 2005).

It is likely that in natural fertility populations, women who have high parity also have the longest reproductive life spans. Early menarche and late menopause increase the total time during which a woman can become pregnant. Furthermore, early menarche and late menopause, and a high number of children should all be positively related. One can expect to see all these characteristics present in women who have good nutritional status.

Good nutritional status is related to early reproductive maturation, which by itself adds a few years to the total time during which a woman can get pregnant. Early maturing girls have higher levels of reproductive hormones in their cycles than girls who mature later, and this difference persists for years past menarche (Vihko and Apter 1984). If they are in good nutritional status for most of their adult life, women have cycles with high levels of hormones. As discussed previously, high levels of hormones lead to an increased chance of conception and thus to increased parity. Therefore, in women who are in good nutritional condition, a negative effect of high parity on bone density may not be detectable because it is counterbalanced by the effect of having many high-estrogen cycles.

A study of Amish women, who are characterized by very high parity (7.6 live births on average), showed that mothers with more children had higher mineral density in their hip bones in later life (Streeten et al. 2005). However, the positive effect of parity on bone structure was no longer statistically significant when the BMI of the woman was taken into account. In this group of Amish women, a positive trend of increasing BMI with increasing parity was detected. That is, the women who had the most children also had a high BMI. This relationship between parity and BMI is common among women from industrialized countries.

A high BMI usually results from deposits of body fat. As mentioned earlier, fat tissue is the most significant source of estrogens in postmeno-pausal women. Adrenal glands secrete an androgen steroid hormone called androstenedione, which is converted into estrogen by the enzyme aroma-tase in the adipose tissue. Overweight women have more adipose tissue, and therefore higher levels of estrogen in postmenopausal years. Higher estrogen levels mean stronger bones. High-parity Amish women have a later age at menopause and higher cumulative estrogen exposure (calcu-lated as age at menopause minus the age at menarche) than women with lower parity (Streeten et al. 2005). This indicates that high-parity Amish women have more estrogen during both premenopausal and postmeno-pausal life. High exposure to estrogen is likely helpful in counteracting the draining effect of reproduction on maternal bone density.

In addition, women in industrialized countries have rather low fertil-ity. A large study in Italy with over 40,000 participants did not show any effects of having children on bone density, but it compared women hav-ing no pregnancies with women who only had one or two children (De Aloysio et al. 2002). As pointed out by Ann Prentice (2000), women who were never pregnant are not the best reference group for these kinds of studies. They may suffer from a variety of physiological problems, in-cluding infertility and reduced levels of estrogen. It is, therefore, too early to conclude that numerous pregnancies and lactations have no effect on maternal bone health.

In developed countries, pregnant women are advised to take calcium supplements. However, some evidence suggests that there are physio-logical adaptations during pregnancy and lactation that provide the calcium needed for the developing fetus and for milk production without additional dietary supplementation. Changes in calcium metabolism and bone metabolism occur in early pregnancy even before the fetal demands for calcium begin. From an evolutionary perspective, a woman should be able to use calcium from her skeletal reserves for all subsequent preg-nancies and lactations without needing additional supplementation—opportunities for supplementing dietary calcium intake did not exist for our ancestral hunter-gatherers. However, some postreproductive bone loss should be expected in women with relatively poor nutritional status and many children, especially if the pregnancies are closely spaced.

Conducting studies of women from populations with poor nutritional status is complicated. Diets in such countries are often not only low in calcium but also in other nutrients, and poor women usually do not have sufficient intake of protein. Lactating women from the Gambia who had diets very low in calcium (less than 300 milligrams per day; by comparison lactating women in the United Kingdom consumed almost 1,200 milligrams of calcium per day) received a calcium supplementation of 700 milligrams per day. After three months, however, calcium absorption did not differ between supplemented and unsupplemented mothers (Fairweather-Tait et al. 1995). In addition, calcium supplementation did not have any effects on bone metabolism (Prentice et al. 1998). It is well known that diets that are low in protein impair intestinal calcium absorption, and individuals who consume a low-protein diet have a lower bone mineral density (Kerstetter, O'Brien, and Insogna 2003). It is unclear whether the diet of the Gambian women, perhaps due to low protein content, was a cause of their low calcium absorption from the supplements.

Calcium supplementation in pregnancy and lactation, therefore, seems not to be needed. Women who take calcium supplements during pregnancy and lactation do not differ in the magnitude of loss of bone mineral density from unsupplemented women. In fact, in well-nourished women an additional intake of calcium may even be detrimental for health because it is associated with a higher risk of kidney stones and urinary tract infections as well as reduced absorption of iron and zinc (Apicella and Sobota 1990; Hallberg et al. 1992; Prentice 2000). In summary, more studies on women with high parity are needed before we can understand the relationship between reproduction, bone health, and the risk of osteoporosis.

Immune Function

It has been suggested that reproduction may negatively impact immune function and increase vulnerability to disease, especially infectious diseases (McDade 2005). No direct data support the existence of trade-offs between the costs of reproduction and immune function in human females. We may suspect that immune function is indeed compromised by reproductive processes. Pregnancy involves many immunological changes that serve to protect the developing fetus from rejection by the maternal immune system (Fleischman and Fessler 2011). These changes are not neutral for the maternal organism, and pregnant women are at increased risk for bacterial and viral infections, including increased susceptibility to human immunodeficiency virus (HIV) infection (Hoff 1999).

High reproductive effort may also contribute to accelerated immunosenescence and early aging (McDade 2005). Reproduction is costly, and each reproductive event elicits changes in the maternal physiology and metabolism. Some of these changes may be temporary, but in women of high fertility they may accumulate and lead to long-term damage. Especially vulnerable to paying the costs of reproduction are women living in poor environments: such mothers rely on many physiological adaptations to support the costs of pregnancy and lactations. But employing these adaptations is not without detrimental consequences for the maternal organism.

Are high costs of reproduction associated with ultimate trade-offs? If each reproductive event is associated with oxidative damage, decline in immune resistance, and other negative physiological and metabolic changes, women with high fertility should experience higher morbidity (the evidence for which was reviewed in Chapter 5) and, in consequence, higher mortality. In other words, do women with many children live shorter lives? Chapter 7 addresses this topic.

7

THE ULTIMATE TEST OF THE

COSTS OF REPRODUCTION

Life Span

Reproduction and Longevity: Depleted Mothers

The relationship between reproduction and longevity has been the focus of many epidemiological and historical demographic studies. The epidemiological research has most often been concerned with the differences in disease and mortality risk between women who have or do not have children, rather than with the relationship between the number of children and a woman's life span. This is understandable because these studies are usually conducted in modern populations from developed countries where women have low fertility, which excludes women with high parity. The value of these studies lies in their collection of information about many potentially confounding—and thus interesting—modern lifestyle characteristics such as smoking, drinking, and body weight. By contrast, information about such lifestyle characteristics is usually not available in the studies that examine historical demographic data, but the latter have the advantage of including women who had a large number of children, who therefore bore significant costs of reproduction.

Very few of the studies with an interest in maternal fertility and life span have addressed the question of whether a similar relationship might also exist for fathers. The most likely reason for fathers having been neglected is because they do not bear the direct, physiological costs of reproduction. However, including fathers is crucial because such studies could generate explanations for the mechanisms responsible for the phenomena observed in mothers. Thus, if a study were to show that both maternal and paternal longevity is negatively affected by the number of children, that finding would suggest that costs other than those associated with pregnancy and lactation must be considered, such as the psychological stress related to supporting a large family, or poor nutritional status from sharing limited resources with many family members, or the complexity of interfamily interactions. Such findings might also suggest that, in the studied population, people with large families have higher mortality due to lower economic status and thus poor health. If the mortality rates of both mothers and fathers were similar, it would mean that the physiological costs borne by women have little additional impact. Including fathers, therefore, serves as an important test of a hypothesis that reduced life span in mothers with high parity is not merely a result of socioeconomic causes.

In a contemporary Israeli population, the lowest mortality risk was documented for women with two children (Manor et al. 2000). By comparison, women with no children and those with more than two experienced higher mortality (Manor et al. 2000). A similar, U-shaped relationship between the number of children and maternal life span has been described in several other studies (Green, Beral, and Moser 1988; Lund, Arnesen, and Borgan 1990; Manor et al. 2000).

In a historical Swedish population (1766–1885), the number of children ever born by a woman had a negative impact on maternal longevity (Dribe 2004). Giving birth to four or more children increased maternal mortality by 30 to 50 percent in comparison with women who had fewer children. Having four or five children instead of none or one shortened the woman's life span by 3.5 years. Further analyses revealed differences among four groups of farmers who lived in this rural population. These groups differed in economic status, and it turned out that the negative relationship between fertility and mortality was restricted only to landless women. For women from families of crown or noble tenants, or for the semilandless who had a better socioeconomic status, reproductive costs were not detrimental. Landless women who suffered higher parity-related mortality had a lifestyle that required high levels of physical labor. Mortality among the men (as this was one of the few studies that analyzed this relationship) was not significantly affected by the number of children. This suggests that the relationship described for women results from the physiological and energetic costs of reproduction rather than their socioeconomic status or from some particular characteristics of the lifestyle of multichild families that increased the risk of parental mortality independent of parity.

In a historical population of northwest Germany (1720–1870), the number of children had a negative impact on longevity but again only among the poor, landless women (Lycett, Dunbar, and Voland 2000). What is remarkable is that a *positive* relationship between the number of children and longevity was discovered among the women of higher economic status. Thus, trade-offs between the costs of reproduction and longevity applied only to women when reproduction represented a nontrivial part of their overall energy budgets.

Historical data from the British peerage, which clearly characterize women with energetically sufficient diets and low workloads, nevertheless have suggested a negative relationship between fertility and longevity— but only after including an additional variable of "unobservable health" (Doblhammer and Oeppen 2003), which has provoked some criticism (Gavrilova and Gavrilov 2005).

In the natural fertility population of the Old Order Amish from Pennsylvania in the eighteenth and nineteenth centuries, who had an average fertility of 7.2 children, a statistically significant negative impact of parity on life span was found only for women who had more than fourteen children (McArdle et al. 2006). Parity from one to fourteen was associated with an increase in maternal life span. After adjusting for the number of children, late age at last birth was strongly associated with an increase in the life span. Each additional year of age at last birth was associated with an average increase in life span of 0.29 years. The paternal life span also increased with the number of children.

How do we explain the results of the latter study? Beeton, Yule, and Pearson (1900) tested the relationship between reproduction and longevity in the belief that the fittest individuals should have the most offspring. The authors of Amish study provided a similar "superior individuals" explanation (McArdle et al. 2006). Parents with high fertility and longer life spans may represent a fraction of the population where, due to genetic or developmental backgrounds, both fertility and health are superior. Late age at last birth may be a marker of longer-lasting reproductive potential, possibly a later age of menopause and, perhaps, a marker of a slower rate of overall aging.

On the other hand, the finding of the same positive relationship between the number of children and longevity for both mothers and fathers suggests that socioeconomic factors may play a role. The benefits provided by children in the Amish community, where family ties are very important, may outweigh the physiological costs born by mothers, except for those with ultrahigh parity. Maternal nutritional status and duration of lactation were not discussed in that study.

Because having sons is thought to be more energetically expensive for mothers than having daughters, the gender of children is also an impor tant variable to consider when exploring a link between energetic and physiological costs of reproduction and longevity. Boys have a faster rate of intrauterine growth and a heavier average size at birth; given their larger body size, they may have higher lactational demands (Powe, Knott, and Conklin-Brittain 2010). Among the Finnish Sami (Helle, Lummaa, and Jokela 2002) and in a Flemish village (Van de Putte, Matthijs, and Vlietinck 2003), having sons decreased maternal life span while having daughters did not.

In four small Polish agricultural villages, however, analyses of parish demographic records collected by Ilona Nenko for the period of 1886 to 2002 showed that both the number of sons and the number of daughters decreased maternal life span, and did so to the same degree (Jasienska, Nenko, and Jasienski 2006). Each son or daughter born to a mother decreased her life span by as much as ninety-five weeks—almost two years—on average. To explain why the impact on the life span is so pronounced, one should remember that, in addition to costs of reproductive events and poor nutrition, women must endure the intense physical labor typical of rural life. Anna, whose life story in the previous chapter was used as a model for analyzing the costs of reproduction in rural farmers, came from one of the villages that participated in this research.

Daddy's Daughters and His Life Span

Data from Polish villages showed quite an unexpected relationship between the number of children and the longevity of fathers. The total number of children, just like in most other studied populations, did not have a significant effect on paternal longevity. When the effects of sons and daughters were analyzed separately, however, the number of sons did not affect the longevity of their fathers, but the number of daughters did. The effect was positive: each daughter increased paternal longevity by seventy-four weeks! Such an effect on parental life span has not been demonstrated in other studies, but, as noted before, few studies have addressed this issue at all.

Why is it beneficial for fathers to have daughters? Historical data do not allow us to answer this question. However, there are a few possible although not mutually exclusive explanations of this phenomenon. The studied rural population in Poland had, and still has, a traditional, patrilineal model of family structure. The oldest son inherits property and land after his father. This traditional rule of inheritance is almost always followed, and thus sons do not need to compete much for paternal approval. Daughters receive money or other assets (for example, cows or pigs) when they get married, but there is no rule that "the oldest daughter gets all" or more than other daughters. Therefore, competition among daughters for paternal favor is expected, as the one who is most favored by her father may gain more resources. Daughters compete by taking care of their father; since the daughters are involved in cooking and preparing food, fathers with more daughters most likely have better nutritional status. In addition, fathers with more daughters may live in more hygienic conditions because daughters clean the house and do the laundry.

Some studies show that men who have children have lower levels of testosterone (Gray et al. 2002; Gray and Campbell 2009; Gettler et al. 2011), which may be beneficial for their health because testosterone acts as an immunosuppressor (Muehlenbein and Bribiescas 2005). It is not known, however, if gender composition among children affects the testosterone levels of their fathers. Perhaps families with more sons experience a higher level of male competition, which may require fathers to maintain higher levels of testosterone, while the fathers with more daughters do not bear these kinds of costs (Jasienska, Nenko, and Jasienski 2006).

More on Mothers and Longevity: Confounding Factors

Returning to the question of whether reproduction is detrimental for female longevity, a number of studies of historical populations have documented a positive association between the number of children and longevity. Similarly, in a French-Canadian cohort of women living in the seventeenth and eighteenth centuries, longevity increased with an increasing number of children, especially for women who were relatively old at last birth (Muller et al. 2002). It has been hypothesized that this may indicate slower ovarian and overall aging (Dribe 2004). But we also reviewed studies that show the opposite relationship between fertility and life span: fertility reduced the maternal life span, which suggests long-term reproductive costs in women. There also were studies that found neither a positive nor a negative relationship between the number of children and the longevity of their mothers (for example, Le Bourg et al. 1993).

It is possible that such conflicting findings are due, at least in part, to the fact that the relationship between fertility and life span is not easy to study. First of all, in populations where people use contraception, the number of children may not be high enough to have an impact on longevity. Second, in affluent populations, the physiological and energetic costs associated with a high number of children may be relatively easy to meet for women, and no trade-offs with life span occur. Finally, the same factors that are related to high fertility may be *independently* related to high mortality. Therefore, we may observe a positive correlation between fertility and mortality, but this relationship may not be causal. In many populations with low economic status, this may be the case. Poor people give birth to many children because contraception is not used, mortality among children is high, and children are viewed as economic assets. The same people, because they are poor, also suffer from high mortality due to poor nutrition and a high risk of infectious diseases. Therefore, there is a positive relationship between fertility and mortality in such cases, but only because in poor people both variables go in the same direction.

Positive Effects of Reproduction

The cumulative costs of pregnancies, as shown previously, are paid even by women who have good nutritional status through a higher risk of cardiovascular disease, diabetes, and stroke. But not every study shows increased mortality in women who have high number of children. Why? Some studies even show an opposite effect: a lowered risk of mortality in women who have given birth to many children. We may speculate about the methodological, social, and biomedical reasons behind discrepancies in the results of these studies, but one aspect is particularly interesting because it illustrates the importance of trade-offs. The very same features of reproductive life that involve the highest metabolic and physiological costs of reproduction—that is, early reproduction and high fertility— may, paradoxically, also have a protective function, leading to decreased mortality from certain diseases.

Young age at first reproductive event and a high number of children are the most important factors protecting women against breast cancer and other reproductive cancers. The protective effects of these factors are well documented (Kvale 1992; Mettlin 1999; Hinkula et al. 2001; MacMahon 2006). Breast cancer risk is also decreased by breastfeeding, which suppresses ovarian activity, although the short-term, nonexclusive (i.e., when maternal milk is supplemented by other foods) lactation commonly practiced by women in economically developed countries may not have as much of a beneficial, protective role (Kvale 1992; MacMahon 2006). However, a study that analyzed results of forty-seven studies from thirty countries (both economically developed and developing) showed that the relative risk of breast cancer decreased by 4.3 percent per each twelve months of breastfeeding (Beral et al. 2002). In women from Nigeria, the risk of breast cancer decreased by 7 percent per each twelve months of breastfeeding (Huo et al. 2008) and in women from India by more than 9 percent (Gajalakshmi 2000).

It may, in fact, be expected that breastfeeding may lead to a more pronounced reduction in breast cancer risk for women from developing countries than for women from economically developed countries. Breast-feeding will cause long-lasting ovarian suppression but only under certain conditions, that is, frequent nursing and poor nutritional status. In !Kung hunter-gatherers, nursing episodes occur as often as every fifteen minutes, and this results in low levels of estradiol and progesterone in mothers (Konner and Worthman 1980). But even such frequent nursing is unlikely to cause long-lasting suppression when the mother is in good nutritional condition (Valeggia and Ellison 2004). For these reasons, women from economically developed countries, even when they are breastfeeding for a long period of time, experience a much earlier resumption of ovarian activity.

Many studies have documented the protective effect of pregnancies against reproductive cancers, but a recent study from Finland is unusu-ally powerful because it included a large number of women of high par-ity and with a reliably diagnosed cause of death (Hinkula et al. 2006). The study analyzed almost 88,000 women with a parity of five or higher (referred to as grand multiparity), out of which 3,678 women had at least ten deliveries. Most high-parity women in this study belonged to the Laestadian movement within the Lutheran church, which prohibits the use of contraception of any kind. The investigators compared the death rates for these women with the average death rates for the Finnish female population. Grand multiparity women had 36 percent lower mortality from breast cancer and 32 percent lower mortality from cancer of the uterus and ovaries. Their mortality from all cancers was reduced by 11 percent.

Date like these strongly suggest that trade-offs between the costs of reproduction and the benefits from reducing the risk of reproductive can-cers are important to consider when attempting to understand the rela-tionship between reproduction and longevity in women. These trade-offs may have a different impact on longevity in different populations. The risk of breast cancer results from high, lifetime exposure to estrogens and

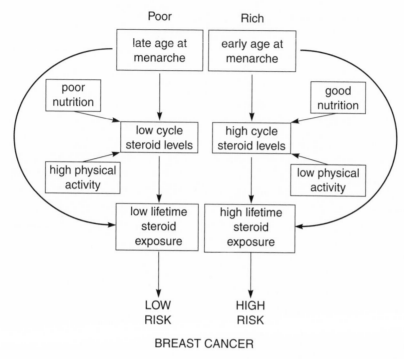

Figure 7.1. Women from poor and rich populations differ in the magnitude of their lifetime exposure to steroid hormones and thus in their risk of breast cancer.

progesterone (Key and Pike 1988; Bernstein 2002). Each reproductive event suppresses menstrual cycles. The higher the number of pregnancies a woman has and the longer cumulative lifetime breastfeeding, the lower the number of menstrual cycles she experiences during her life. If she starts reproduction early, she has additional protection against breast cancer. Pregnancy occurring early in life, by inducing the differentiation of breast tissue, reduces its susceptibility to neoplastic transformation (i.e., the development of tumors). In addition, the postpregnancy period is characterized by low levels of endogenous estrogens, which may further repress potential tumor growth.

In poor agricultural societies, however, women in general have low levels of the steroid hormones estrogen and progesterone produced in menstrual cycles (Figure 7.1). As I discussed in Chapter 1, nutritional shortages and intense labor periodically suppress ovarian function. These same factors, in addition, postpone the age of sexual maturation when

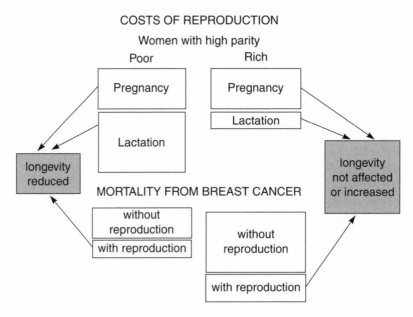

Figure 7.2. Effect of equally high parity on longevity in women from poor and rich historic populations. Women from poor populations had higher costs of reproduction due to the costs of lactation, which usually was not a factor for well-off women. Because the risk of breast cancer in poor women was low to begin with, an additional reduction in that risk due to reproduction had little impact on overall mortality. By contrast, women from rich populations had a high initial risk of breast cancer, so its reduction as a result of reproduction was substantial. In addition, in the latter, the costs of reproduction were easily met by increased intake and decreased physical activity (not shown on the diagram). Thus, high parity in well-off women is not expected to lead to a reduction in mortality and may even increase longevity.

cycles begin. Therefore, these women have a low lifetime exposure to es-trogens. They have fewer cycles, and during each of their cycles relatively low levels of hormones are produced.

In economically well-off populations, women have high levels of hor-mones in menstrual cycles because they rarely experience ovarian suppres-sion. The factors limiting energy availability that cause ovarian suppres-sion are mostly absent from their lives. Many of the historical genealogies analyzed in the context of the impact of fertility on longevity were based on data from well-off families. High fertility in women from such families significantly reduced lifetime exposure to estrogen because each pregnancy completely prevented occurrence of menstrual cycles. In contrast, estrogen exposure in women living in poor conditions (due to poor nutrition and intense physical work) is low to begin with, and high fertility has a much lower relative effect in reducing lifetime estrogen levels. Estrogen levels in these women are low anyway. Therefore, in women from rich environments, high fertility may not translate into reduction in longevity for multiple reasons. The costs of pregnancies are often not accompanied by costs of lactations, or these costs are very low. Costs of pregnancy and lactation are easily met by additional energy intake, especially when there are no energetic demands from physical activity (Figure 7.2).

Furthermore, there are trade-offs between reproductive costs and benefits provided by protection against breast cancer. It is even possible that in well-off women the benefits may outweigh the costs, and that high fertility may, in fact, be related to life span extension. Finnish women with high parity had much higher mortality from all metabolic diseases, especially diabetes and cardiovascular diseases, than the average for the population, but these risks were clearly outweighed by lower mortality from cancers because their *overall* mortality was slightly lower than the population average (Hinkula et al. 2006).

Well-Being, Children, and Grandchildren

It is obvious that children not only impose biological costs but also provide some benefits to their parents. They help in the household in various ways and provide support, which would be been especially important for elderly parents in preindustrial societies in the absence of a pension system or health insurance (Tsuya, Kurosu, and Nakazato 2004). For our discussion here, it is important to compare the costs and benefits of having children, especially from the perspective of parents who are old.

Economic conditions and social determinants, especially family structure and patterns of caring for aging parents, determine how children influence parental longevity and well-being. Having children had a positive effect on parental life expectancy in some historical European communities (Tsuya, Kurosu, and Nakazato 2004). In the Liaoning province in China between 1749 and 1909, having no sons increased the mortality of both parents (Campbell and Lee 2004). Particularly women were affected: their mortality rose by 12 percent if they had no living sons. If, in addition to not having sons, the older men or women were also widowed, their chance of dying was over 50 percent higher than for individuals who had at least one son and a living spouse. For older men, having living sons had no effect on mortality unless the man was a widower.

Traditionally, many older people lived and, in some societies still do, with one of their children and his or her family. Their longevity is thus not only influenced by the costs and benefits arising from the physiological costs paid during their reproductive years or by their relationships with their children but also, in later years, by their relationships with their grandchildren. The costs and benefits of such interactions are important to consider when discussing health and mortality risks. Family structure was an important correlate of mortality in two agricultural villages located in the least economically developed regions of preindustrial Japan between 1716 and 1870 (Tsuya, Kurosu, and Nakazato 2004). Female infants had a lower probability of dying if a grandmother was living in the household, but grandmothers had no significant impact on the mortality of male infants. In contrast, the presence of a grandfather increased the mortality risk for male children by over 40 percent.

The authors of this study explained this negative effect of the grandfather by invoking competition for the same resources between children and the elderly, who both depended on the household's working adults for support (Tsuya, Kurosu, and Nakazato 2004). In patriarchal systems, grandfathers are more serious competitors than grandmothers. Why the grandfather's presence did not affect the mortality of girls is less clear, but it has been suggested that girls are not even entitled to compete for resources and always receive less food than their brothers. In the study mentioned earlier on the demography of the Liaoning province in China, the presence of a grandfather significantly increased the risk of mortality for children aged two to fifteen of both genders, although the effect for girls was much stronger (Campbell and Lee 2004). The presence of other relatives did not have a significant effect on children's mortality, except that having the mother alive decreased mortality for boys but not for girls.

Grandmothers are important, demographically speaking, in several societies. In historical Germany (Voland and Beise 2002) and Japan (Jamison et al. 2002), grandchildren survived at higher rates if they grew up with their maternal grandmother in the household. In the Hadza hunter-gatherers of Tanzania, the nutritional status of grandchildren was better (Hawkes, O'Connell, and Jones 1997), and in the rural Gambia, both the nutritional status and survival of grandchildren were enhanced if grandmothers were present (Sear, Mace, and McGregor 2000).

But what about the grandparents themselves? Much is known about the effect that grandparents have on grandchildren, but do these interactions between older and younger generations have any effect on the grandparents' well-being? We cannot answer this question without considering the social structure of a given society. In patriarchal systems, the costs and benefits of such interactions may be different for grandmothers and grandfathers. Grandfathers usually are not expected to contribute much to the household, and in times of economic hardship they may feel entitled to a higher share of resources than grandmothers. Land or property are often inherited through the male line; therefore, males in old age are treated with more care and respect. The expectations toward grandmothers are much higher: they are often required to become involved in childcare or physical labor. Even in some matrilineal societies, such as the Tonga from the Gwembe Valley of rural Zambia, men can accumulate enough to support themselves well into old age while women are left without any wealth (Cliggett 2005, 65–66).

Furthermore, the costs related to interactions with grandchildren may also vary for maternal versus paternal grandmothers. Maternal grandmothers, according to the concepts of kin selection (Hamilton 1964) and paternity uncertainty (see Russell and Wells 1987), are expected to contribute more toward the children of their daughters than the children of their sons. Because paternity is always associated with some degree of uncertainty, grandmothers cannot be certain that their son's children are related to them, but such problems never exist with the children of their daughters. It is likely then that grandmothers living with their biological daughters will be ready to pay higher costs for the help they are likely to provide. In contrast, grandmothers living with their daughters-in-law should be less driven to provide costly help to grandchildren who may not be genetically related to them. Several studies have documented the benefits for grandchildren living with maternal grandmothers and the lack of such benefits when children live with their paternal grandmothers (Hawkes, O'Connell, and Jones 1997; Hawkes et al. 1998; Sear, Mace, and McGregor 2000; Jamison et al. 2002; Voland and Beise 2002). It is, therefore, very likely that such benefits for children are not cost-free for their grandmothers.

Benefits from Having Grandchildren

A different reasoning assumes the lack of trade-offs between benefits grandmothers provide to grandchildren and the costs for the grandmothers themselves. A lack of trade-offs is postulated by Sorenson Jamison and coauthors (2005) when interpreting the data from a peasant village in Japan from 1603 to 1867. They tested the "grandmother's clock" hypothesis developed by Sarah Blaffer Hrdy (1999), which combines the "good mother" hypothesis with the "grandmother" hypothesis. These hypotheses attempt to explain the surprisingly long postreproductive life span of the human female.

The good mother hypothesis proposes that human females must live long enough to raise their youngest offspring to maturity (Sherman 1998). Given the long period of immaturity of human children, a woman who had her last birth at age forty-five should live to be at least sixty years old to protect her reproductive investment. The central idea behind the grandmother hypothesis is that postreproductive women benefit by helping raise their grandchildren (Hawkes et al. 1998). This benefit should be understood in evolutionary terms and expressed in evolutionarily relevant currency. Grandmothers who provide help increase their own inclusive fitness, where fitness is measured by passing genes to the next generation, which can be achieved either by reproducing or helping relatives who share copies of the same genes to reproduce or survive.

The data from the Japanese villages seem to support both hypotheses (Sorenson Jamison et al. 2005). Women who gave birth to their last child relatively late were less likely to die in their postreproductive years than were women who had their last child relatively early, which supports the good mother hypothesis. Furthermore, grandmothers in this Japanese village were less likely to die than women who did not have any grandchildren: the life span of a woman was extended, on average, by 4.75 years if she had grandchildren. The authors argued that grandmothers were better nourished than nongrandmothers "since caretakers who are responsible for making sure the children are fed have additional food available for themselves, too" (Sorenson Jamison et al. 2005, 106). This explanation assumes that help provided by grandmothers does not render any costs to them but rather provides additional benefits, possibly nutritional, which ultimately result in longer lives.

The researchers assumed that grandmothers provided help to grandchildren, but they were not able to demonstrate that help based on the available data. It is possible that the economic conditions of the studied population were good, and that the grandmothers who helped their grandchildren did not suffer high nutritional or energetic costs and possibly even benefited by maintaining rich social interactions with their extended family. In poor economic conditions, it is unlikely that the help provided by grandmothers to their grandchildren would result in an extension of their own lives in old age. When grandmothers share limited resources with their grandchildren, their own nutritional status is likely to suffer, and "emotional fulfillment" is unlikely to compensate.

It is also possible that the gender composition among grandchildren is an important factor to consider when weighing the extended costs and benefits of reproduction. Gender differences in how children's well-being is affected by social factors are known in many populations around the world. In a comparative study of several European and Asian eighteenth- and nineteenth-century populations, child mortality was found to be affected by the number of people over fifty-five years old living in the household (Alter et al. 2004). These adults were most likely grandparents or, in some cases, other relatives such as unmarried or widowed uncles and aunts. In almost all of these populations, an increase in the number of older family members increased the chances of survival of boys aged two to fourteen but decreased the chances for girls in the same age group. It was suggested that these gender differences imply that girls had to compete for resources within the household yet boys were not only free from such competition but actually benefited from the presence of additional adult family members (Alter et al. 2004).

Gender inequalities among children in resource allocation are visible from analyses of short-term economic stress. Such short-term dietary shortages were clearly important for shaping mortality patterns in historical rural Japan (Tsuya, Kurosu, and Nakazato 2004). Increased prices of rice, the main staple for the population, increased the mortality of female children but had no significant effect on male children. It is likely, therefore, that grandparents can obtain a higher share of limited resources when living with a family having more granddaughters. When, in contrast, grandparents live within a family with a higher number of grandsons, they may face higher competition for resources, with potentially negative impacts on their health and longevity.

Why Childless Women Do Not Have the Longest Life Span

If there are high metabolic costs to reproduction, can we assume that women who have no children should, on average, live longer than women who have children? Two hypotheses suggest that this should be the case. First is the hypothesis stating that the physiological and metabolic costs of reproductive processes cause wear and tear of the maternal organism. The "maternal depletion" hypothesis has been discussed in this chapter and in Chapter 6. The second hypothesis, developed to explain the evolution of human longevity, assumes the action of genes to enhance longevity by diverting resources from reproduction to maintenance and repair (Kirkwood and Holliday 1979; Westendorp and Kirkwood 1998). These genes effectively act to limit reproduction. During periods of energetic restriction, such genes support physiological maintenance instead of supporting reproductive processes (Kirkwood, Kapahi, and Shanley 2000). Both hypotheses generate the same predictions regarding being childless and long-lived. Childless women simply have no metabolic reproductive costs for nongenetic reasons or have alleles that limit reproduction and thus shield them from reproductive costs. Of course, the second hypothesis is difficult to test because such women, in addition to possessing these alleles, also do not bear reproduction-related metabolic costs.

Most studies, however, do not show a longer life span for childless women. The lack of findings supporting their longer life span has been incorrectly used to suggest that the costs of reproduction are not related to longevity (Gavrilova and Gavrilov 2005). In a study of British nobility, the trade-offs between fertility and longevity became significant only after childless women and women with one child were removed from the analysis (Doblhammer and Oeppen 2003). Childless women may not be an appropriate group among whom to study reproduction and life span, just as they are not the best reference group in studies of the physiological costs of reproduction. Historically, it is likely that a significant proportion of married, childless women had health problems that prevented them from having children. Such health problems were not only related to a reduced ability to conceive or carry to term but also contributed to their increased risk of mortality. In addition, many societies ostracize childless women, which may also contribute to their ill health (Poston and Kramer 1983; Poston et al. 1983). In British women, the risk of heart disease was lowest among women who had two children; higher risks were noted for women with higher parity but also those who had no children or only had one (Lawlor et al. 2003).

When Does Reproduction Reduce Life Span?

Although there is convincing evidence that reproduction is costly and related to long-term changes in female physiology, by no means should a negative impact of reproduction on life span always be expected. The energetic and metabolic costs of reproduction cannot be calculated by merely adding children. Women with a similar number of children may vary in the costs of lactation, which can be substantial. Similarly, they may vary in their ability to meet the energetic and physiological costs of reproduction as a result of differences in lifestyle, dietary intake, and physical activity. The high costs of reproduction and its negative effects on some aspects of health may be outweighed by its positive effects on other health dimensions. For some women, intense reproduction will substantially reduce their risk of reproductive cancers, but for other women those risks are already relatively low even when they do not have many children. The indirect costs of reproduction, such as those paid by grandmothers taking care of grandchildren, may also change the ratio of the overall lifetime costs and benefits of reproduction.

A negative relationship between fertility and longevity may, therefore, be expected in women who not only suffer high costs of reproduction but also have a low dietary intake and high levels of physical activity. High fertility is expected to reduce life span especially for women who have low lifetime estrogen levels caused by growing up and living as adults in an energy-poor environment. For such women, an additional reduction in the level of ovarian hormones (estrogen and progesterone), usually caused by reproduction, is relatively insignificant. High fertility exposes them to a higher risk of cardiovascular disease, adult onset diabetes, and osteoporosis and does not protect them from hormone-dependent cancers.

How we as biological organisms deal with energy has been one of the main themes of this book. From the very beginning of fetal life—and even before that—the availability of energy influences the future biological quality and health of the individual organism. For women, energy is a major determinant of the functioning of their reproductive physiology, with a profound impact on the menstrual cycle, conception, pregnancy, and lactation. By influencing levels of steroid hormones and body weight, energy indirectly affects the risk of many diseases, including breast cancer. Chapter 8 discusses obtaining energy from the diet. It also discusses dietary nutrients and how sufficient, diverse nutrients are as important as having enough energy for many aspects of physiology and health.

In people from industrialized countries, a lack of energy in the diet is no longer a problem. Modern diets are, in fact, often bad for us because they bring excessive amounts of energy with poorly balanced intake of many other components. What we eat today is very different from the dietary spectrum consumed by our ancestors. How much do modern diets differ from the diets that humans had during our evolutionary past? What was the composition of the so-called Paleolithic diet, and how do we know that? Would the Paleolithic diet be still healthy for humans today? Can we expect that adopting this pattern of consumption would lead to a significant reduction in the incidence of modern metabolic diseases, especially cardiovascular diseases, adult onset diabetes, and cancers?

8

EVOLUTIONARY PAST AND MODERN DIET

The connection between diet and human health and disease has been one of the most intensely studied areas of epidemiology and public health. Yet there is still no agreement as to what is the most appropriate diet for humans to consume. Moreover, no other area of study of such profound importance to health seems to produce so many conflicting and contradictory take-home messages. Even the recommendations about such basic matters as the correct proportions of major macronutrients (fats, carbohydrates, and proteins) in our daily dietary intake are still under debate. In the United States and several other countries, many people attempting to lose weight or just trying to "eat healthily" have recently changed their diets from the government-recommended "low fat, high carbohydrate" to "low carbohydrate, high fat" diets. The U.S. Department of Agriculture's official food pyramid—the recommended intake of foods from different food groups—has been blamed for contributing to the obesity epidemic and criticized by some scientists working in the area of human nutrition (Willett and Stampfer 2003). The pyramid has undergone major changes, and new guidance was released in 2005. Recently, nutritional pyramids have been abandoned and replaced by the graphically much simpler MyPlate.

There are debates about the correct proportions of macronutrients in our diet, and there is also no agreement about the recommended doses of the micronutrients: vitamins and minerals. Different types of food become hyped as the most healthy to eat and as providing protection against various diseases, but they quickly lose importance and are replaced by others. Diets from various countries and populations have been used as examples of good and bad, with the so-called Mediterranean diet and U.S. fast food as the epitomes of correct and incorrect diets. But even the Mediterranean diet, although associated with a lower risk of cardiovascular disease than the typical U.S. diet, represents a relatively recent pattern of dietary consumption for our species. The Mediterranean diet is agriculturally derived and thus almost by definition cannot be best suited for human nutritional needs. These needs have been shaped by natural selection for generations, over many thousands of years before people even began to grow food. Recently, there has been a movement promoting pre-agricultural diets, called Stone Age or Paleolithic (Eaton and Cordain 1997; Eaton, Eaton III, and Konner 1999; Milton 2000; Cordain et al. 2005). This trend is grounded in evolutionary thinking and deserves to be discussed in more detail.

Craving for "Unhealthy" Food

It takes a long time before evolutionary adaptations become established in a population in response to a new environment (Stearns, Nesse, and Haig 2008). It has been argued that human physiology and metabolism are best suited to our past Paleolithic environment, including the diet (Eaton, Eaton III, and Konner 1999; Cordain et al. 2005). A frequently used and very suggestive example illustrating the poor fit between human physiology and modern lifestyle is the case of our craving for "unhealthy" foods. The modern diet, which is high in fat and simple sugars, is blamed for obesity and the high incidence of many diseases, including cardiovascular diseases and non-insulin-dependent diabetes, the plagues of industrial populations.

Dietary changes leading to a reduced intake of sweet and fat foods are very difficult to implement because most humans crave such tastes. This is not surprising as these tastes have always served as the most trustworthy indicators that the food has a high energy content. Energy was in limited supply in the environment of our evolutionary past, and therefore individuals who craved and actively sought such energy-dense foods were in better nutritional shape then those who preferred the taste of energy-poor foods such as leaves or roots. At the same time, there was no danger of overeating because foods with a high fat and sugar content were quite rare in the Paleolithic environment. The meat of wild animals has a much lower fat content than the meat of farm animals (O'Dea 1991), and the only foods with a high sugar content in our evolutionary past were wild honey, rather difficult to find, and ripe fruits, available only seasonally for short periods of time.

The rarity of high-fat, high-sugar foods in the human evolutionary environment is unfortunate for us. Had they been more common, it is likely that appropriate physiological adaptations would have evolved to allow increased consumption without negative health consequences. Our evolutionary legacy can be blamed for the fact that today many people in industrialized countries who have easy access to unlimited food resources are left with a taste for fat and sugar but lack the physiology to handle it without negative health consequences.

The Paleolithic-type diet, matched to our evolved physiology and metabolism, seems like an appropriate recommendation for modern humans. Promoting a Paleolithic diet is, however, not free from problems. First, many critics stress that there was no such a thing as a *universal* Paleolithic diet. During human evolutionary times, even though all humans lived as hunters and gatherers, incredible variation in food consumption existed both over time and between various regions (Ungar and Teaford 2002). Such variation implies that humans were forced to adapt to a wide variety of dietary environments, and the expected result was a broad plasticity in human dietary needs. Our distant ancestors more than 4 million years ago evolved the ability to consume a wide variety of foods, as suggested by the changes in postcranial and dental morphology (Ungar and Teaford 2002).

Fortunately, although there clearly was great variability in evolutionary diets, some common patterns emerged. It is certain, for example, that no society consumed high amounts of simple carbohydrates (i.e., sugars). All societies consumed animal protein, often in high quantities. Putting aside Arctic populations, meat was mostly very lean compared with the meat of domesticated animals consumed today, and thus consumption of saturated fats was rather low. Diets included a large variety of plants and a very high intake of vitamins and minerals.

Paleolithic Diets in Modern Environments

Paleolithic diets worked well in Paleolithic environments. Modern humans, however, may have different dietary needs than their ancestors owing to differences in lifestyles, mainly with respect to physical activity and reproductive patterns. It is obvious, for example, that the caloric intake appropriate for a hunter-gatherer would be much too high for a sedentary office worker. The same can be true for certain nutrients. Low levels of physical activity, and in women lower nutritional requirements resulting from a reduced reproductive output, may mean that today the nutritional requirements for vitamins and minerals are lower than they were for our human ancestors.

Because humans constantly change their environment, adaptations that worked in past environments are not necessarily optimal for the current environment. As Richard Lewontin remarked, "No two successive generations would be selected in the same direction and for the same characteristics. In general the organism and the environment must track each other continuously" (Lewontin 2000, 126).

Promoting Paleolithic diets as appropriate for contemporary humans is also problematic when one considers evolutionarily derived trade-offs. Paleolithic diets were clearly suitable for promoting reproduction and survival during reproductive years. Therefore, such diets contributed to the good health of young people of prereproductive and reproductive age. But we already know this much: what is advantageous when using evolutionary currency is not necessarily advantageous when health is the currency, especially the health of older individuals. Cravings for any food or nutrient that had the potential to increase reproduction should and must have been promoted by natural selection, even if such physiological and behavioral tendencies were detrimental for health in older age.

It is also likely that in postreproductive years a different diet is optimal because physiology changes as people age. It can be assumed, therefore, that the Paleolithic diet probably was not the best nutritional choice for Paleolithic older people and cannot possibly be today. However, an evolutionary ideal of the best diet for older age may not exist at all: natural selection does not operate during postreproductive years, or if it does, then only indirectly. Grandparents, and especially grandmothers, in some societies seem to have a positive impact on the health and survival of their grandchildren (Hawkes, O'Connell, and Jones 1997; Sear, Mace, and McGregor 2000), as discussed earlier in Chapter 7, and thus may still be exposed to some selective pressures.

How Do We Know What Paleolithic People Had on Their "Plates"?

Paleolithic diets are very difficult, almost impossible, to reconstruct as the remains of food items, especially from plants, are rarely preserved, particularly for hunting and gathering societies that did not have permanent houses. One way to reconstruct the Paleolithic diet is to study the diets of contemporary hunter-gatherers. Although this approach clearly provides important insights, several methodological problems emerge.

Few populations remain today that still support themselves exclusively by hunting and gathering. Most data collected by ethnographic and anthropological studies during the twentieth century when these populations still had their traditional lifestyles usually do not allow for quantitative dietary analyses. Furthermore, modern hunter-gatherers often live in marginal environments, usually in areas that are not suitable for agriculture; thus, they live in environments different, usually poorer, than those occupied by their ancestors. It is likely that the modern environments of hunter-gatherers have a different combination of foods available for consumption than had Paleolithic environments. Moreover, recent decades have witnessed the dramatic disappearance of the hunting-gathering way of life, as many foragers have moved to permanent settlements, either forced by governmental policies or voluntarily seeking a more modern lifestyle and more frequent contact with the outside world.

Despite all the methodological challenges of research in this area, a conclusion might be drawn that Paleolithic diets are, most likely, much more appropriate for modern humans than other diets, especially diets that are agriculturally based. If so, why does their importance not emerge in epidemiological research? The Mediterranean diet, especially the traditional diet of poor rural populations from this region, and the Japanese diet are frequently shown to be associated with a lower risk of metabolic diseases, and such findings influence public health recommendations. But the health benefits of Paleolithic diets are almost never discussed by epidemiologists. One reason for this state of affairs immediately comes to mind, namely, that epidemiological findings are mostly based on studies with sample sizes large enough for different causes of morbidity and mortality to be analyzed. Unfortunately, Paleolithic diets are not a popular topic of research, and thus they are not even considered by the organizations that provide recommendations about healthy lifestyles.

Evolutionary Patterns of Consumption and Modern Dietary Recommendations

Knowledge about Paleolithic diets and the evolutionary pattern of food consumption is important not only for finding the optimal diet but also for understanding why following dietary recommendations is so difficult for modern people. First, as described before, our preferred tastes are not for foods that in the modern environment are most healthy. On the contrary, people who follow their innate food preferences mostly eat fats and sugars. Obviously, and tragically, this is true in many societies with unlimited access to such foods.

Paleolithic patterns of food consumption also suggest that while humans have some physiological mechanisms limiting the amount of food they are able to eat (Morrison 2008), these limits are set very high, as suggested by observations from Australian Aborigines. Australian Aborigines, when leading their traditional hunting and gathering way of life, were known to eat 2 to 3 kilograms of meat during one long meal (O'Dea 1991). However, occasions when such a bountiful amount of meat was available were rare, usually only when a kangaroo was killed; once that happened, people took maximum advantage of the situation. It has been suggested that the ability to consume a lot of food during these infrequent "feasts" was critically important for the survival of hunter-gatherers. A surplus of energy, stored as fat, provides crucial energy reserves during periods of food shortages. In Paleolithic environments, people most likely stored little food, and most was consumed shortly after it was gathered or hunted. Storing food was difficult or impractical for several reasons, including the risk of it being stolen or a necessity to share it (Wrangham et al. 1999).

This tendency to eat all available food in one sitting was not lost with the Australian Aborigines' change from a traditional to a modern lifestyle. Store-bought meat, usually cheap fatty beef or lamb, can still be consumed by westernized Aborigines in very large quantities (O'Dea 1991)—but with one very important difference. Three kilograms of lean meat, such as kangaroo, provided about 3,000 kilocalories. This seems like a lot, but it pales by comparison with the same amount of meat from domesticated animals, which yields up to 12,000 kilocalories! A similar pattern of the "old way" of consumption applies to other foods. Eggs, purchased by the dozen, are usually cooked by the Aborigines all at once and may be consumed by a single person (O'Dea 1991).

The importance of eating regular meals, avoiding snacking, and beginning the day with a substantial breakfast are often essential aspects of modern nutritional recommendations. Such recommendations, however, have little to do with the evolutionary pattern of human consumption. People ate whenever food was available. Like many other hunter-gatherers, Australian Aborigines ate one main meal during the day, in the late afternoon after having returned to camp with food gathered or hunted that day (O'Dea 1991). Snacking during the day was a typical component of the evolutionary dietary consumption. Grubs, fruit, gum, ants, and honey from wild bees were eaten throughout the day. Shellfish and fish were also eaten whenever caught—sometimes cooked, at other times raw. When a kangaroo was killed, hunters sometimes would cook and eat its liver before carrying the meat to the camp. Food consumption varied also a lot from day to day, or across the seasons. There were days with limited intake when plants were mostly consumed, and "feasts" that occurred when a large animal was killed.

Many people today prefer a steak or burger to broccoli. This preference for meat in the diet is not new for our species. In one of the earliest studies of the nutrition of !Kung San hunter-gatherers living in the Kalahari desert in Africa, their caloric intake was estimated to be 2,140 kilocalories per day, and their protein consumption was an average of 93.1 grams per day (Lee 1968). Meat contributed 33 percent of the calories in their daily diet, and vegetable foods were 67 percent. This dietary intake was judged to be adequate, based on a recommended daily allowance of 1,975 kilocalories and 60 grams of protein for a hypothetical person of !Kung San stature and high levels of physical activity. (Note that there was no information provided as to whether this dietary assessment was for males, females, or an average of both genders.) The !Kung San dietary approach was summarized as followed: "the Bushmen of the Dobe area eat as much vegetable food as they need, and as much meat as they can" (Lee 1968).

Gathered vegetable food is always available for hunter-gatherers, but game animals are relatively scarce, difficult to find and catch. Putting meat on the plate requires much more effort. Among the !Kung, one man-hour of hunting yields only about 100 kilocalories, whereas one hour (mostly done by women) of gathering yields about 240 kilocalories. If people had not evolved a taste for meat, would they be spending so much energy hunting game animals? Most likely not. If our ancestors from the evolutionary past considered meat as appealing as the !Kung do today, we can assume that humans have an evolutionarily based tendency to value meat much more than any plant food. Chimpanzees, our close primate relatives, also hunt and clearly enjoy eating meat (Stanford 1998). Is this one of the reasons that most of us do not eat enough vegetables? The most likely answer is yes.

Paleolithic Diets

In Australia, the Aborigines lived exclusively as hunters and gatherers beginning at least 40,000 to 50,000 years ago until about 200 years ago, when Australia was colonized by the Europeans (O'Dea 1991). Small groups of Aborigines in the remote regions continued to live as foragers as recently as the 1970s. Their diet consisted of a broad variety of animal and plant foods. Not only mammals and birds were consumed, but also reptiles, insects, and marine species. Meat was an important component of their diet, but wild animals have low fat content, and the main type of fat obtained from animal food comprised long-chain polyunsaturated fatty acids (Naughton, O'Day, and Sinclair 1986).

The sources of foods were different for hunter-gatherers than they are for us today. For example, plant foods in the Aboriginal diet were derived from tuberous roots, seeds, fruits, nuts, gums, and nectars (O'Dea 1991). Wild plants, in comparison with cultivated varieties, are richer in protein, fiber, potassium, magnesium, calcium, and vitamins. Wild plum *(Terminalia fernandiana),* part of the Aboriginal diet in northern Australia, has a vitamin C content of 2 to 3 percent of wet mass, which is higher than in any other known food. Carbohydrates in wild plants have lower rates of digestion and absorption than those in cultivated varieties, so consumption of these carbohydrates does not lead to rapid jumps in the concentration of glucose in blood—a feature important for maintaining low insulin levels (Englyst and Englyst 2005).

A Model of Paleolithic Diet Composition

Eaton and Konner (1985) proposed a model for evaluating Paleolithic nutrition based on analyzing the nutritional content of wild game and uncultivated plants, examining archeological remains, and studying dietary data from contemporary foragers. Recent analyses based on their model have resulted in the best assessment of the human evolutionary diet to date (Eaton, Eaton III, and Konner 1997). The average Paleolithic daily diet is thought to be based on about 900 grams of meat and about 1,700 grams of plant food and provided 3,000 kilocalories. The diet had 37 percent protein, 22 percent fat, and 41 percent carbohydrates (calculated as the percentages of energy consumed) (Figure 8.1).

Another Paleolithic model proposed the following ranges of the consumption of macronutrients: 19 to 35 percent protein, 28 to 58 percent fat, and 22 to 40 percent carbohydrates (Cordain et al. 2000). Adults in the United States obtain, on average, about 15 percent of their dietary energy from protein, 34 percent from fat, 49 percent from carbohydrates, and an additional 3 percent from alcohol. In their model, Eaton, Eaton

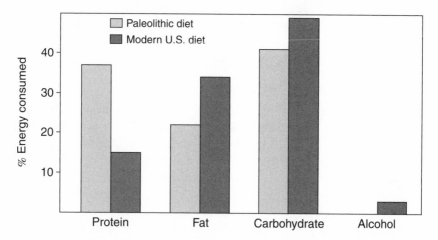

Figure 8.1. Differences in dietary intake of macronutrients in the Paleolithic diet and in the modern U.S. diet. Alcohol consumption contributes additional energy to the modern U.S. diet. Data from Eaton, Eaton III, and Konner (1997) and Cordain et al. (2000).

III, and Konner (1997) estimated that the daily Paleolithic intake of all analyzed vitamins and minerals was considerably higher than either the U.S. government's recommended dietary allowance or the actual consumed values by Americans. The Paleolithic intake of vitamins was, for each 1,000 kilocalories consumed, between 1.5 times higher (for folate) to 8.4 times higher (for vitamin C) than in the modern United States. The intake of minerals was between 1.7 times higher (for calcium) to 5.8 times higher (for iron). The only mineral consumed in higher quantities in the modern United States is sodium, with an intake of 1,882 mg (milligrams) in the modern U.S. diet versus only 256 mg in the Paleolithic diet, again calculated for each 1,000 kilocalories. When it comes to macronutrients in the diet, not only their percentages differ when comparing the Paleolithic diet and current consumption, but equally important are differences in the food sources from which the carbohydrates, fats, and protein originate (Eaton, Eaton III, and Konner 1997). In Paleolithic diets, most of the carbohydrate intake came from fruits and vegetables, and very little from cereal grains or honey. In the modern U.S. diet, only 23 percent of carbohydrates are consumed as fruits or vegetables, and in Europe their consumption is even lower. The rest come mostly from refined flour, sugars and sweeteners that are very low on nutrients, dense in energy, and quickly cause elevations in blood glucose levels (Englyst and Englyst 2005).

Fats and Cholesterol

There are two main types of polyunsaturated fatty acids: omega-3 and omega-6. Their relative dietary proportions seem to be very important for many aspects of the metabolism. Omega-3 fatty acids are found in fish oil, while omega-6 fatty acids are obtained from sources such as corn and safflower oils. Excessive levels of omega-6 fatty acids relative to those of omega-3 fatty acids are associated with an increased risk of disease, including cardiovascular diseases and probably breast, colorectal, and pancreatic cancers (Bagga et al. 1997; Lands 2005; Hibbeln et al. 2006; Geelen et al. 2007; Funahashi et al. 2008). The relative proportion of these two types of fatty acids was much healthier in the hunter-gatherer diet than in most modern diets. In the Australian Aborigines, the relative proportion of omega-3 and omega-6 polyunsaturated fatty acids was about 1:3, but in the Western diet this proportion is about 1:12 (Sinclair and O'Dea 1990). We need more omega-3 in our diets.

As mentioned previously, wild animals are lean; in the past, saturated fat contributed only 6 percent to the average dietary energy intake. Current recommendations from the American Heart Association limit intake of such fats to less than 7 percent (Lichtenstein et al. 2006). Given the high proportion of meat consumption in the Paleolithic diet, the daily intake of cholesterol was indeed quite high at 480 mg, whereas today an intake of 300 mg per day or less is considered desirable (Lichtenstein et al. 2006). Despite their high dietary intake of cholesterol, modern hunter-gatherers had very low blood cholesterol levels of about 125 milligrams per deciliter (mg/dL). There has been a slight decrease in the U.S. population's total serum cholesterol levels since 1988, but between 1999 and 2002 the average levels were still very high at 203 mg/dL, particularly with high levels of low-density lipoprotein cholesterol, the "bad" kind (Rosamond et al. 2007).

Total serum cholesterol concentrations are made up by a few different types of lipids. Because cholesterol molecules cannot move by themselves inside the body, they require transport from cell to cell. This transport is provided by lipoproteins: low-density lipoprotein (LDL), the so-called bad cholesterol, and high-density lipoprotein (HDL), or "good" cholesterol. Having too much of the LDL cholesterol is a problem because it has a tendency to build up on the inner wall of arteries; together with other molecules, it eventually can form into plaque that narrows the arteries. Narrowed arteries easily can be blocked by a blood clot; when blood cannot pass to the heart or brain, this may cause a heart attack or stroke. By contrast, HDL cholesterol usually carries cholesterol away from the arteries and may even remove cholesterol from plaque that has already begun to build (American Heart Association 2012). Having insufficient good cholesterol is a potential problem because its deficiency increases the risk of heart attack.

The low total serum cholesterol of hunter-gatherers in spite of their high cholesterol intake may derive from the relative proportion of different types of fat in their diet (Eaton, Eaton III, and Konner 1997). Their consumption of polyunsaturated fats exceeds 1.4 times the consumption of saturated fats. In the modern American diet, this pattern is reversed: Americans consume 2.5 times more saturated fats than unsaturated fats. Two saturated fatty acids, C_{14} myristic and C_{16} palmitic, can raise total serum cholesterol levels. Both C_{14} and C_{16} acids are present in meat, but wild animals have very low quantities of either. In addition to the beneficial proportion of fats in their diet, the low cholesterol level in hunter-gatherers likely also is related to their high levels of physical activity. Low levels of total cholesterol and LDL cholesterol and high levels of HDL cholesterol are often found in people who are physically active (Owens et al. 1990; Murphy et al. 2002).

Serum cholesterol levels can be also high due to consumption of trans fatty acids, even though they are mostly monounsaturated. Most trans fatty acids are made by the food industry through partially hydrogenating plant oils, so their consumption in the Paleolithic diet was practically nonexistent. Because trans fatty acids are mostly used in fast foods, fried foods, and baked goods, they are presently consumed in high amounts. Trans fatty acids raise serum cholesterol levels in a manner similar to that of C_{14} and C_{16} fatty acids, but in addition trans fats also reduce levels of the beneficial HDL cholesterol fraction (Judd et al. 1994). Not surprisingly, consumption of trans fatty acids increases the risk of cardiovascular disease: a 2 percent increase in energy intake from trans fatty acids is associated with a 23 percent increase in the incidence of the disease, and may also increase the risk of diabetes (Mozaffarian et al. 2006).

Hopefully, consumption of trans fats will decrease in the near future, at least in some locations. In 2003, Denmark introduced a law that banned the sale of all products containing trans fats. Similar though less restrictive bans exist in several cities in the United States, including New York City where, in 2006, the Board of Health banned the use of trans fats in all restaurant foods. In many other locations, however, people may not even be aware that they eat trans fats because there are no requirements to list them on food labels.

Protein and Fiber

Out of the three macronutrients (fat, carbohydrates, and protein), the intake of protein in the Paleolithic diet diverges most from today's recommended values. The recommended intake is a value of 0.8 to 1.6 grams/ kilogram per day, while the values estimated for Paleolithic humans were more than twice as high at 2.5 to 3.5 grams/kilogram per day. Eaton, Eaton III, and Konner (1997) point out that in other primates that consume a predominantly plant-based diet, including our close relatives the chimpanzees and gorillas, the intake of protein is also very high, ranging between 1.6 and 5.9 grams/kilogram per day. Plant food can have high protein content, and in addition primates are known to selectively choose food with higher protein content. For example, gorillas from Gabon in Central Africa select leaves that are high in protein and low in fiber compared with the generally available vegetation (Rogers et al. 1990).

Protein in the diets of hunter-gatherers comes mostly from consumption of meat, often in substantial quantities. Over 70 percent of all hunter-gatherer societies derived more than half of their dietary energy from meat, and only 14 percent derived more than half from plant foods (Cordain et al. 2000).

The high protein content of modern diets may be linked to an increased risk of several diseases, but it is important to realize that the high protein intake today comes mostly from eating domesticated meat and dairy products, which are linked to a high consumption of saturated fat. In the Paleolithic diet, this unfavorable association did not exist because the meat was lean and a lot of protein came from plant sources.

The consumption of fiber in the Paleolithic diet also greatly exceeded the amount of fiber consumed today. The high consumption of plant food by hunter-gatherers contributes to a fiber intake of more than 100 grams per day (Eaton, Eaton III, and Konner 1997). In contrast, modern Americans consume less than 20 grams per day. The American Heart Association recommends a fiber consumption of 14 grams for each 1,000 kilocalories per day, well below the values common during human evolutionary history.

Consumption of fiber may be of special importance for women because eating a lot of fiber appears to reduce the risk of breast cancer (Gerber 1998). In premenopausal women from the United Kingdom, those who consumed more than 30 grams/day of fiber had about a 50 percent lower risk of breast cancer than women who consumed less than 20 grams of fiber per day (Cade, Burley, and Greenwood 2007). This reduction in risk is most likely due to the lower circulating levels of estrogens and progesterone found in women who consume fiber-rich diets (Kaneda et al. 1997; Rock et al. 2004). As discussed before, high levels of these hormones promote development and growth of cancer. In a study that carefully controlled the intake of all nutrients, an increase in dietary fiber consumption to 40 grams per day resulted in a significant reduction in the level of estrogens during the menstrual cycle's follicular phase (Goldin, Woods, and Spiegelman 1994). Another study showed a 10 to 20 percent reduction in serum estrogen levels in premenopausal women who took wheat bran supplements for two months and had high total intake of fiber of 32 grams per day (Rose, Lubin, and Connolly 1997).

A high-fiber diet also may reduce the level of estrogen in postmenopausal women. In women with a history of breast cancer who took part in a dietary intervention study, a diet with high intake of fiber (29 grams per day on average) and low intake of fat led to a significant reduction in the estradiol concentration after one year (Rock et al. 2004). Statistical analysis showed that the reduction in estradiol was due to the high intake of fiber and not to the low intake of fat.

Paleolithic Diets in Modern People

As we have seen, the Paleolithic diet had a much higher intake of protein, fiber, and many vitamins and minerals than is recommended by nutritionists today as optimal for human consumption. Are the nutritionists wrong? One way to interpret this discrepancy may lead to the conclusion that if all those nutrients were consumed today in quantities resembling the Paleolithic diet it would be detrimental for the health of modern people. Another view is that nutritionists vastly underestimate our physiological and metabolic needs for such nutrients. At present, it is hard to choose one of these assumptions over the other.

The apparently high Paleolithic consumption of vitamins and minerals is still far below the minimum toxic doses (Eaton, Eaton III, and Konner 1997). No convincing evidence supports the notion that consumption of protein or fiber at the levels estimated for Paleolithic humans would be harmful for human health today. On the other hand, the consumption of certain macronutrients or micronutrients should not be analyzed outside of the context of the Paleolithic lifestyle. Currently, low physical activity and, for women, low reproductive costs may have lowered not only our energy requirements (i.e., caloric intake) but also our requirements for specific nutrients. For example, calcium supplementation in women who were well nourished but had a presupplementary dietary intake well below the Paleolithic values was associated with a higher risk of kidney stones, urinary tract infections, and reduced absorption of iron and zinc (Apicella and Sobota 1990; Hallberg et al. 1992; Prentice 2000).

Humans living a modern lifestyle may also benefit from foods that were not frequently part of the Paleolithic diets, such as alcohol. Many studies have shown that moderate alcohol consumption reduces the risk of cardiovascular diseases, probably by increasing coronary flow and arterial dilation (Gazzieri et al. 2006). But women cannot drink alcohol without some trade-offs. Even a moderate consumption of alcohol to decrease the risk of heart disease leads to a higher risk of breast cancer (Michels et al. 2007).

Other products unknown during the Paleolithic era also may provide health benefits in modern populations. Green tea is full of antioxidants, which have cardioprotective and cancer-fighting abilities. Tea was not part of the Paleolithic diet; rather, the diet contained many other plants unavailable today that had a high antioxidant content. Plants that grow under intense sun exposure use such chemicals to protect their cells from the free radicals produced during photosynthesis (Benzie 2003). Many modern plants, especially if they are grown in greenhouses, have far lower antioxidant concentrations than wild plants or plants that grow in regions with intense, long exposure to sunlight such as the Mediterranean. Perhaps this also contributes to the lower incidence of metabolic diseases in people from the Mediterranean regions.

There is clearly a need to evaluate the health consequences of Paleolithic diets in humans who live an industrialized, Western lifestyle to confirm that what worked in Paleolithic environments will work equally well in modern times. Toddlers are very picky when it comes to food, and they often cannot be convinced to eat leafy vegetables such as spinach or they insist on eating just one food type at a time. Elisabeth Cashdan (1998) argued that these behaviors have deep evolutionary roots. Plant leaves

184

are often toxic, and thus avoiding such foods reduces the risk of poisoning. Eating one food at a time allows for efficient learning about the consequences of ingesting a novel food item. Toddlers' eating habits were perhaps adaptive in ancestral environments, but it is not clear how these choices influence the nutritional status of children in modern, urban societies.

It can also be expected that optimal diets based on the Paleolithic model should be qualitatively different for men and women, even after adjusting for differences in body weight and composition. The differences between the Paleolithic and modern lifestyles for both genders involve diet and physical activity, but for women major changes also have occurred in reproductive patterns. An earlier age at menarche, more menstrual cycles during the lifetime, higher levels of hormones produced during their cycles, fewer children, shorter breastfeeding periods, and easily met costs of reproduction may be related to different dietary needs than those of our female Paleolithic ancestors.

In addition, the dietary intake that humans had in the Paleolithic era may not necessarily be the most beneficial for health and survival. Here, I must make an unexpected taxonomic leap to make a point. A fascinating experimental study of field crickets *(Teleogryllus commodus)* illustrated that the effects of nutrition are sex specific (Maklakov et al. 2008). Crickets were assigned to receive diets with varying combinations of nutrients. As it turned out, the diets with low protein and high carbohydrate concentrations (a 1:3 protein:carbohydrate diet, i.e., three times as much carbohydrates as protein) maximized the life span of both males and females. However, this diet was not the best choice for having top reproductive performance, at least not for all crickets. The maximum reproductive performance was achieved by each gender on a different diet. In males, the diet that maximized life span also maximized the most important component of their reproductive performance, the frequency of their calling songs. But in female crickets, the daily egg-laying rate and lifetime egg output were maximized when they were fed a diet with an equal concentration of protein and carbohydrates (a 1:1 protein:carbohydrate diet)— very different from the diet most beneficial for survival. This study clearly shows that female crickets cannot choose a diet that will allow them both high reproductive performance and high survival.

We do not have equally convincing experimental data for human females to show that optimal diets for reproduction are perhaps not the best for health. But with respect to iron, human females face similar nutritional dilemmas as female crickets. In a pregnant woman, an adequate intake of iron is important for fetal development. Anemic mothers are at increased risk of premature delivery and of having a baby with low birth weight, low body fat, and behavioral developmental issues (Milman 2006). However, a high concentration of iron in the maternal organism increases oxidative stress (which is already high in pregnancy): high levels of free radicals damage maternal cells (Casanueva and Viteri 2003), which may lead to faster aging (Rattan 2006).

Oxidative stress is only one of the problems resulting from an iron-rich diet. Immune functions in pregnant women are greatly impaired, as can be concluded from their higher rates of new infections and increased severity of persistent infections such as malaria (Thong et al. 1973; Kochar et al. 1998). Pathogens need iron for proliferation and thrive in iron-rich internal environments (Fessler 2002). Therefore, a diet rich in iron is beneficial for fetal development but at the same time may put the mother at higher risk. Lower (but not anemic) concentrations of iron in the diet may maximize the mother's survival (especially for women living in environments with high exposure to pathogens), but higher concentrations may be better for her reproductive performance.

The Health of Hunter-Gatherers

A review of the health status of hunter-gatherers based on data available for several populations suggests that malnutrition happened only rarely among people living traditional lifestyles (Dunn 1968). Their daily diet was judged to be adequate for meeting their basic nutritional requirements. Periods of starvation were common among agriculturalists but were rare for hunter-gatherers living in tropical and temperate regions, although they were more common among Arctic populations, especially during the winter. Chronic diseases appear to have been relatively rare in populations of hunter-gatherers, and people were exposed to few risk factors causing such diseases.

Kerin O'Dea (1991) reviewed data from Australia showing that Aborigines had a low body mass index (a BMI not higher than 20kg/m^2) and low resting blood pressure. Nor did these parameters follow the Western pattern of increasing with age. The fasting glucose levels and cholesterol concentrations were also lower among the hunter-gatherers compared with the urbanized Aborigines or Australians of European descent.

186

The positive relationship between the hunter-gatherer way of life and their health was documented by an Australian study on "reverse lifestyle change" (O'Dea 1984). But before Kerin O'Dea performed that study, another "experiment" on lifestyle and health had already been conducted by life itself: the health condition of many Aborigines rapidly declined after they adopted a westernized lifestyle. The changes in their diet and reduction in their physical activity resulted in a high incidence of obesity, with 50 to 80 percent of adults over thirty-five years of age being overweight or obese (O'Dea 1991). The prevalence of diabetes in Aborigines with a Western lifestyle is ten times higher than in Australians of European descent. Coronary heart disease and hypertension also are much more common among westernized Aborigines.

The "reverse lifestyle change" experiment involved returning to the hunting and gathering life, albeit only for a short period (O'Dea 1984). In the 1980s, many Aborigines living a Western life still had the knowledge and skills of hunting and gathering, so they were able to resume the old, traditional way of subsistence. Ten diabetic, overweight, middle-aged Aborigines from the Mowanjum Community in Western Australia agreed to take part in the study. Remarkable changes occurred to the health condition of these people after just seven weeks of hunting and gathering. They lost 8 kilograms of weight, on average, showed improvements in all metabolic symptoms of diabetes, and had a reduction in the factors known to increase the risk of heart disease.

Agriculture and the Human Biological Condition

Lifestyle-related health problems did not originate in recent times with the Industrial Revolution, urbanization, and unlimited access to energy-dense food products, although all these changes have contributed to poor health. A negative impact of lifestyle was seen much earlier in human history. Many people think that the transition from foraging to agriculture was beneficial and improved the quality of human life. Agriculture has often been seen as the entry to civilization. People did not have to move around, and could live in permanent houses and accumulate possessions. Permanent settlements allowed the storage of harvested foods for use when times were hard. Work became more seasonal, so the overall demand for physical labor decreased. Populations increased in size, suggesting increased fertility, decreased mortality, or both, which has been used as the main evidence that life conditions improved.

187

The last several decades, however, have brought new evidence that has changed our view of the agricultural transition. Agriculture was indeed related to changes in the human biological condition, but most of those changes were for the worse (Larsen 2002). The diet in general changed from one with a very high variety of foods to a reliance on a few cultivated crops, rarely supplemented by foraging. Isotope analyses of skeletal remains suggest that some early farmers, even those living in proximity to the seacoast, did not supplement their agricultural diets with seafoods (Papathanasiou, Larsen, and Norr 2000). It is estimated that the !Kung hunter-gatherers eat about 105 species of plants and 144 species of animals (Lee 1984), and Australian North Queensland Aborigines exploit about 240 species of plants and 120 species of animals (Gould 1981). The diet of an average person living in an agricultural society has just four species of plants and two species of animals (Bogin 2001, 158).

Not only did dietary variety decrease with agriculture, but the main dietary staples provided a poor nutritional base. The consumption of maize, for example, reduces the bioavailability of iron, and maize is also deficient in essential amino acids, which contributes to poor growth (Larsen 2002). Millet, one of the main African staples, and wheat, a main staple in Europe and temperate Asia, are also deficient in iron when milled. Rice is deficient in protein. Skeletal remains indeed suggest a high prevalence of iron-deficient anemia in agricultural Neolithic populations.

A high consumption of plant carbohydrates has been tied to the widespread decline in oral health, as documented by an increase in dental caries (Larsen 1995). In North America, for example, pre-agricultural populations had less than 5 percent carious teeth, but agricultural populations consuming maize had over 15 percent (Larsen 2002). Farmers, in comparison with foragers, had a higher frequency of teeth enamel defects, resulting either from malnutrition or infectious diseases, or a combined impact of both factors (Larsen 2002). Nonspecific bony lesions, called periosteal reactions, which indicate the occurrence of infectious diseases, increased in frequency as well after the agricultural transition.

The prevalence of infectious diseases was believed to increase due to sedentism, crowding, and poor sanitary conditions. Poor dietary quality, especially a decreased availability of protein, was reflected in the reduced rate of growth during childhood and low final adult height (Cohen and Armelagos 1984). Comparison of foragers with farmers shows much greater robustness of bones before agriculture: the size and structure of the bone tissue is shaped in response to mechanical demands, and more

active populations have larger and more robust bones than more sedentary people (Larsen 2002).

An evolutionary approach to diet together with knowledge about human physiology, genetics, and metabolic processes allows us to understand
at least two issues: the difference between evolved patterns of consumption and modern diets, and the negative consequences for our health of
eating food that is unsuitable for our evolved requirements. Clearly, we are
far from identifying the perfect diet for modern people. The reasons for
that, as discussed in this chapter, are complex, ranging from the unavail-
ability of original Paleolithic food sources to the possible incompatibility
of the Paleolithic diets with our modern lifestyle. The perfect diet may not
even exist at all. The dietary patterns that led to high reproductive success
were certainly promoted by natural selection even while causing, in the
long run, damage to our physiology and an increased risk of diseases in
old age.

Besides diet, physical activity is clearly the most important lifestyle-
related predictor of health. Just as knowing about a Paleolithic diet may
help us find out what we should eat today, knowledge about Paleolithic
physical activity may also be useful for designing recommendations for
beneficial types of exercise. But similarly there are some problems with
relying on a Paleolithic-based recipe with respect to physical activity.

Is jogging beneficial for women? The answer is yes. Was jogging a frequently used mode of locomotion among Paleolithic women? Unlikely.
Walking is clearly more practical when gathering food, and it is an easier
way of moving around during pregnancy and when carrying a child. Does
that mean that women today should choose walking instead of jogging
to achieve the health benefits of physical activity? If we were to base our
recommendations for modern women on the Paleolithic model, we would
answer yes to this question. But this recommendation would not be very
practical for most modern women—if they were to try to follow the Paleolithic model, they would have to do a *lot* of walking.

9

EVOLUTION AND PHYSICAL ACTIVITY

In contemporary !Kung San hunter-gatherers, an average adult walks about 2,400 kilometers per year, amounting to about 6.6 kilometers per day. An average !Kung male has a day range of 14.9 kilometers, compared with an average U.S. male office worker whose walking day range is not much higher than zero kilometers (Cordain, Gotshall, and Eaton 1997). !Kung women walk these long distances while carrying substantial loads. A child is carried 1,500 kilometers during its first two years of life. Weighing 3 kilograms when born and up to 15 kilograms by age four, a child usually is carried by the mother on gathering trips in addition to equipment and, on the way back, heavy loads of nuts, roots, and other plants (Lee 1979, 310; Bentley 1985).

Human physiology and metabolism were shaped by the ancestral environment, which forced a certain pattern of physical activity. The exact amount of activity and its daily energetic costs are, of course, unknown and can only be estimated. Just as in the case of human ancestral dietary patterns, studies of hunting and gathering societies provide important insights. The same problems, however, that are associated with deciphering the Paleolithic diet based on studies of contemporary foragers apply to attempts to reconstruct Paleolithic patterns of physical activity. It is clear that in the hunting-gathering way of life both finding the food and preparing it for consumption typically involve physical activity that is either intense or sustained for a long time. For Australian Aborigines, the main activities involve walking long distances, digging in rocky ground for tubers, digging for reptiles, honey ants, and witchetty grubs, and grinding seeds. Although some food is eaten fresh and raw, much has to be cooked (O'Dea 1991; Wrangham et al. 1999; Wrangham and Conklin-Brittain 2003). Cooking large animals requires digging pits and collecting a large supply of firewood, which is also needed to keep people warm, especially during the night. In general, the hunter-gatherer way of life includes activities that are both aerobic and requiring strength (Eaton and Eaton III 2003).

We Have Evolved to Be Active

Undoubtedly there is variation in the level of physical activity among different hunter-gatherer populations, but it is clear that their activity level is much higher than found today among people who have an urbanized lifestyle (Cordain, Gotshall, and Eaton 1997).

There are several ways to express the energy expenditure of an individual, but for comparative purposes the *physical activity level* (PAL), a ratio of total energy expenditure over basal metabolism, is frequently recommended (Panter-Brick 2002). Because it takes into account differences in body size, PAL allows a comparison of the physical activity of different individuals. Body size is an important predictor of the energetic costs of basal metabolism, which is expressed as either the *basal metabolic rate* (BMR) or the *resting metabolic rate* (RMR). Larger individuals have higher rates of basal metabolism.

Modern hunter-gatherers substantially differ in body size. An average !Kung male weighs only 46 kilograms, and an average male in another hunter-gatherer society—the Ache of eastern Paraguay (Hill and Hurtado 1996)—weighs almost 60 kilograms (Leonard and Robertson 1992). PAL values are available for only a few of hunter-gatherer populations: for males they are 1.71 for !Kung, 2.15 for Ache, and 2.2 for Igloolik Eskimo (Leonard and Robertson 1992; Katzmarzyk et al. 1994).

Women have lower levels of physical activity, even when controlling for their smaller body size. The women's PAL values are 1.51 for !Kung, 1.88 for Ache, and 1.8 for Igloolik Eskimo. Based on these values, the activity levels of Ache and Igloolik can be classified as heavy, and that of !Kung as light for women and moderate for men (FAO/WHO/UNU 1985). The PAL values of modern, sedentary people are much lower: 1.18 for men and 1.16 for women (Heyward 1991).

In promoting the evolutionarily based need for exercise in modern humans, Cordain, Gotshall, and Eaton (1997) suggested that the PAL value of 1.87, which represents the mean value estimated for hominids since the appearance of *Homo erectus* (Leonard and Robertson 1992) roughly 1.8 million years ago (Anton 2003), "represents the activity level for which our species is genetically adapted." Such a level of physical activity is not easy to attain by augmenting our generally sedentary lifestyle with casual exercise. A hunter-gatherer male expended 19.6 to 24.7 kilocalories/kilogram each day in physical activity, whereas a sedentary office worker expends just 4.4 kilocalories/kilogram per day (Cordain, Gotshall, and Eaton 1997). If that sedentary office worker would walk 3 miles each day, his activity would increase to only 8.7 kilocalories/kilogram per day, still far below the activity level characteristic of the traditional lifestyle. Running for one hour each day at a speed of about 12 kilometers/hour would be needed for the office worker to match the energy expenditure of human ancestors (Cordain, Gotshall, and Eaton 1997).

Very few people can achieve such high levels of exercise, but even if this were possible, the question remains as to whether running at high speed for 60 minutes then spending the rest of the day sitting at a desk, in a car, and in front of a television still would be equivalent to the daily activities of hunters and gatherers. A long, daily, intense run would result in a comparably high amount of daily energy expenditure, but the types and pattern of overall physical activity would be very different. Does it matter? Is it enough just to match the energy expenditure of hunter-gatherers? To maintain good health, must one follow similar types of physical activities distributed during the day in a similar temporal pattern?

Duration and Patterns of Physical Activity

The American Heart Association recommends that "all adults accumulate thirty minutes of physical activity most days of the week" and suggests that additional benefits would likely be derived if the activity levels exceeded this minimum recommendation (Lichtenstein et al. 2006). At least sixty minutes of physical activity most days of the week is recommended for children and for adults who are attempting to lose weight or maintain a recent weight loss. Moreover, physical activity can be "accumulated throughout the day."

In accordance with these not very demanding recommendations, some epidemiological evidence confirms that any physical activity is better than none. Just a bit of exercise provided by walking for one hour per week seems to reduce the risk of cardiovascular diseases (Oguma and Shinoda-Tagawa 2004). In studies such as the Women's Health Study, the duration of walking was found to be more important than the intensity (Lee et al. 2001); in others, such as the Health Professional Follow-Up

Study, which enrolled about 51,000 men, the intensity and not the duration seemed to be more beneficial for health (Tanasescu et al. 2002). In the latter study, walking pace was associated with a reduced risk of cardiovascular disease independent of the number of walking hours. A half-hour or more per day of brisk walking was associated with an 18 percent reduction in risk.

It is clear, however, that more intense or longer lasting physical activity is associated with increased health benefits, correlating with greater reduction in the risk of cardiovascular disease, diabetes, or breast cancer. However, in accordance with the "too much of a good thing" principle, some studies have shown that such benefits diminish with very intense exercise that risks injury or sudden cardiac death (Bucksch and Schlicht 2006).

The health benefits of light physical activity are not always confirmed by research. The Harvard Alumni Health Study of over 13,000 men failed to find that light activities (activities that raise the RMR less than four times) were associated with reduced mortality rates (Lee and Paffenbarger 2000). Only moderate activities (which raise the RMR four to six times) and especially vigorous activities (which raise the RMR six times or more) led to lower mortality rates. These data clearly show the need of incorporating at least moderate intensity activity into the daily routine and indicate that vigorous activity is the most beneficial. Here are a couple of examples of the energy costs of various activities: walking the dog raises the RMR three times, bicycling to work four times, walking while playing golf 4.5 times, and playing tennis five times. More intense activities include shoveling snow at six times the RMR, jogging at seven times, or running at eight to eighteen times, depending on the speed.

It is important to distinguish between the two main outcomes that are evaluated by studies of physical activity. First, they evaluate the effect of activity on endurance fitness, usually assessed by the maximal oxygen uptake (VO_2max), which reflects the maximal capacity of an organism to transport and use oxygen during exercise of increasing intensity. Second, they study the effect of activity on changing the risk of various diseases and overall mortality (Hardman 1999).

Although engaging in intense activity, usually sports, does improve fitness and is also related to a lower risk of disease, less intense activity may not have an effect on endurance fitness but often (but not always) is related to a reduced disease risk. The American College of Sports Medicine's previous recommendations for physical activity were fifteen to sixty minutes of continuous moderate to vigorous exercise, three to five days per week (Macfarlane, Taylor, and Cuddihy 2006). In the 1990s, in response to the low adherence to their guidelines, the ACSM instead recommended that people should accumulate thirty or more minutes of moderately intense physical activity on most days of the week (American College of Sports Medicine 2005).

Although such recommendations clearly seem to be easier to follow for most people, surprisingly little is known about whether such activity is really beneficial. In particular, the question of whether multiple-bout activity is comparable with longer, single-bout activity has not been addressed by a large number of studies. It can be assumed that the effects of these two patterns of activity should be similar, if fitness and health depend on the total amount of physical activity per day or on the total amount of energy expended in such activities. If so, then a single 30-minute exercise should have equally positive results as three 10-minute exercise sessions spread throughout the day, given that the total energy expenditure of both regimes is the same. Can it be so simply additive? We are not sure how the human organism integrates the information about the amount of spent energy.

To address this question, researchers assigned sedentary adults in Hong Kong to two exercise programs: thirty minutes of continuous activity three to four days per week, or a six-minute activity five times per day, four to five days per week (Macfarlane, Taylor, and Cuddihy 2006). After eight weeks, both groups showed significant improvement in VO_2 max, although the group with one bout of exercise per day improved more. In another study, the increase in VO_2 max was higher in the group with continuous exercise than in the group with three bouts of exercise, but the heart rate decreased to a similar extent in both groups (DeBusk et al. 1990). In another example, walking for thirty minutes per day for eight weeks either in a single bout or in three 10-minute bouts significantly and in a similar way improved the VO_2 max of participants (Osei-Tutu and Campagna 2005). However, in comparison with the nonactive control group, those who did a single bout of walking noticed a greater reduction in their percentage of body fat, their tension-anxiety, and their total mood disturbance, and they also reported increased vigor. In obese, sedentary women, both long-bout and short-bout groups experienced improvements in VO_2max (higher in the long-bout regime), but exercise in short bouts made it easier for the participants to follow the exercise program (Jakicic et al. 1995).

Of course, not only exercise but also habitual activities are related to reduced disease risk, and these activities are clearly not performed in a single bout. For example, stair climbing—a multiple-bout activity—has been associated with reduced mortality risk in men (Paffenbarger et al. 1993). Those who climbed less than twenty flights of stairs per week had a 23 percent higher risk of premature death than those who more frequently took the stairs instead of elevators.

Light activity, mostly resulting from walking, when added to a daily sedentary lifestyle reduces the risk of cardiovascular diseases and adult-onset diabetes. Risk reduction of some other diseases, however, requires more intense activity; although it has been demonstrated that the risk of colon and breast cancer can be reduced by physical activity (McTiernan 2008), at least moderate activity is needed (Thune and Furberg 2001). In general, a so-called dose–response relationship between physical activity and risk of most diseases is suggested: light physical activity may lead to some reduction in disease risk, but higher activity is needed for more substantial benefits. It is understandable that given the overall high inactivity level in many populations, public health recommendations call for only a little physical activity. Demonstrating that such activity may improve health and reduce risk of mortality is clearly important, and such recommendations are thought to be more effective in convincing people to be active. In particular, doing just a little exercise may be more feasible for people who are obese, not very fit, and not healthy. But for most healthy adults, following the advice of just walking for thirty minutes per day is not enough to significantly improve fitness and health. Although very few people will be able to match the activity levels of our human ancestors, clearly an intermediate level of activity should be recommended. People should be made aware of the benefits of higher levels of physical activity than just walking and encouraged to follow more intense physical activity regimes. In fact, the current, nondemanding recommendations from the American College of Sports Medicine may be doing a disservice to healthy adults, advising them that relatively low levels of activity are enough to keep them fit and healthy.

Contemporary hunter-gatherers are in better physical condition than age-matched people from urbanized societies. Hunter-gatherer males aged twenty to forty-nine years have, on the average, a VO_2max of 57.2 milliliters/kilogram per minute, while urban males have only 37.2 milliliters/kilogram per minute (Cordain et al. 1998). Hunter-gatherers are also about 20 percent stronger in leg extension tests (Eaton and Eaton III

2003). A relationship between aerobic condition and reduced risk of several diseases, especially cardiovascular disease, is well established, and with increasing aerobic fitness a higher reduction in risk is observed. Exercise capacity is a good predictor of overall mortality both in healthy people and those afflicted with various diseases (Oga et al. 2003; Church et al. 2004). Moreover, exercise capacity seems to be a more powerful predictor of mortality among men than many other established risk factors of cardiovascular disease (Myers et al. 2002), including being overweight (McAuley et al. 2007).

Burning Fat When Not Active

A physically more active lifestyle contributes to health even when the individual is temporarily less active. A permanent increase in the BMR frequently occurs as a result of sports participation (Poehlman and Horton 1989; Van Zant 1992; Burke, Bullough, and Melby 1993; Sjodin et al. 1996; Tremblay et al. 1997; Dolezal and Potteiger 1998; Morio et al. 1998). A higher BMR means that the individual requires more energy to support metabolic processes also during periods of inactivity; therefore, individuals may, with less effort, be able to maintain a lean body. Increases in the BMR should, in general, be expected in individuals who, due to exercise, have experienced changes in body composition, especially when longer-term physical training has led to increased muscle mass. Muscle, as a metabolically active tissue, is energetically expensive to support (McArdle, Katch, and Katch 1986), and an increased BMR reflects this additional expense.

However, not every training regime produces changes in body composition, and as a result a change in the BMR is unlikely to occur. For example, fifteen weeks of either low- or high-intensity strength training failed to cause significant changes in the body composition of participating women (Taaffle et al. 1995). Not surprisingly, the BMR of these women remained at their pre-exercise level. In contrast to this finding, studies of endurance athletes (who usually have a higher percentage of fat-free mass than less athletic people) and studies during which training caused changes in body composition often have reported positive effects of training on the BMR. In women rowers, changes in BMR reflected changes in fat-free mass (McCargar et al. 1993). Elite endurance athletes had a BMR 16 percent higher than in the nonathletic control group (Sjodin et al. 1996). Significant increases in BMR after ten weeks of training were reported in men engaged in resistance (strength) training, but not in men participating in endurance training—running and/or jogging (Dolezal and Potteiger 1998), possibly reflecting the differences in muscle mass gain.

Not only physical activity but also the rate of energy flow are positively correlated with basal metabolism (Burke, Bullough, and Melby 1993; Bullough et al. 1995). A high rate of energy flow occurs when an individual has both high energy intake and high energy expenditure. One study reported that RMR rose as a result of training but only in men characterized by a high rate of energy flow (Bullough et al. 1995). In women, the RMR also showed a positive relationship with energy flow and aerobic fitness (determined by VO_2max) (Burke, Bullough, and Melby 1993). Long-term physical activity, especially when associated with changes in body composition and high rates of energy flow, may thus induce significant and beneficial increases in the basal metabolism.

Why There Is No Such Thing as the Optimal Dose of Physical Activity

Epidemiological studies of physical activity seem to yield much more uniform results than do studies of diet. There is general agreement that physical activity reduces the risk of many diseases and lowers the overall risk of mortality. However, the agreement does not extend to what types of activity or which kinds of exercise are most beneficial. Furthermore, epidemiological studies have attempted to find an optimal amount of exercise and a threshold below which exercise starts or stops being beneficial. Although it is understandable that identifying an optimal dose would be very practical for promoting physical activity, the notion of an optimal dose does not make sense for several reasons.

First, let us consider the following dilemma. The risks of breast cancer and osteoporosis are both reduced by physical activity, but the relationship between the amount of activity and risk reduction is expected to be very different for each of these diseases (Figure 9.1). Physical activity reduces breast cancer risk because it reduces estrogen levels (evidence for this was reviewed in Chapter 1), but it reduces risk of osteoporosis *despite* reduced estrogen levels. Mineral bone density is maintained, among other factors, by the action of estrogen. Low estrogen levels are, therefore, detrimental to bone density and are related to an increased risk of osteoporosis. Physical activity works as a preventive mechanism against

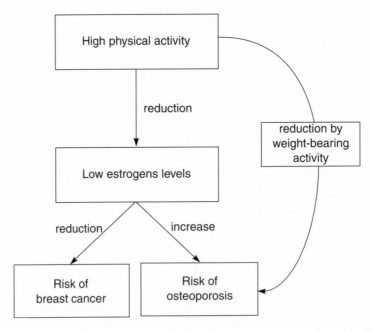

Figure 9.1. Physical activity lowers the risk of breast cancer by reducing the level of estrogen. Unless the activity has a weight-bearing component, it may *increase* the risk of osteoporosis because low estrogen levels can contribute to low mineralization of bones.

osteoporosis mostly because of the positive weight-bearing effect it has on bones. In premenopausal women, very intense, long-duration physical activity that dramatically lowers estrogen levels could be beneficial in terms of breast cancer prevention yet increase the risk of osteoporosis (Keen and Drinkwater 1997). This would be true especially when the physical activity is not weight-bearing but of a different type—for example, swimming. Here we encounter yet another trade-off: the same factor, estrogen, reduces the risk of one disease but leads to an increased risk of another disease. The same lifestyle factor, physical activity, lowers the risk of one disease but may increase the risk of another.

Second, individual variation complicates the protective effect of exercise against breast cancer. The reason that the same dose of physical activity cannot be prescribed for every woman to lower the risk of breast cancer lies in differences among individuals in the sensitivity of their ovarian response. We have reason to believe, as discussed earlier, that such sensitivity is related to conditions experienced during fetal development. The ovarian function of women whose developmental conditions were good, as indicated by their relatively large size at birth, has a lower sensitivity to physical activity during adulthood. Women who were born as smaller babies have a higher sensitivity.

There are important implications of these findings for breast cancer prevention. The minimal amount of physical activity that is expected to reduce the risk of this disease will vary for different women, depending on their birth sizes. Women born as small babies react with reduced estrogen levels to moderate physical activity, but the ovarian function of women born as large babies does not respond to that same activity. These women need activity of higher intensity or longer duration to attain a similar, steroid-reducing beneficial result. It should be noted that currently this relationship between size at birth and the level of physical activity needed to reduce breast cancer risk is only hypothetical. Epidemiological studies are needed to confirm that this is really the case.

Is Physical Activity an Adaptation to Keep Diseases Away?

During human evolutionary history, physical activity was a salient feature of the lifestyle. Just as in the case of diet, however, natural selection designed a good fit between intense physical activity and physiology for individuals in their reproductive years. The postreproductive years, when most diseases occur today, most likely were not the target of selective improvements. The protective effect of physical activity on the risk of breast cancer is not a direct product of natural selection but rather a by-product of the evolved response of ovarian function to environmental conditions (Figure 9.2). Intense physical activity suppresses ovarian function—and therefore estrogen production—to prevent pregnancy during times when too much energy is needed for activity and not enough is left for reproduction.

Reproductive ecologists believe that this ovarian response is adaptive and that it evolved by natural selection. However, the effect of physical activity on reduction of breast cancer risk is not an adaptation. The protective effect of activity results from a combination of two different processes: an adaptive ability of estrogen-producing processes to respond to physical activity and an adaptive ability of estrogen to increase the rate of breast-cell division. An increase in the number of cells is necessary for lactation, but it also increases the probability of cancer. Reduction of the

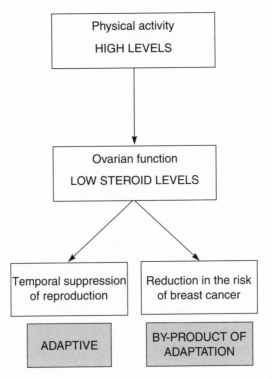

Figure 9.2. Physical activity reduces the risk of breast cancer, but this relationship does not reflect adaptation. It is a by-product of an adaptive response of ovarian function that reacts to intense physical activity by suppression. The resulting low levels of ovarian steroids, especially when they persist for a long time, are related to the reduced risk of breast cancer.

risk of breast cancer by physical activity is, therefore, just a by-product of evolutionary adaptations as well as a lucky coincidence.

What Activity and How Much of It?

The relationships among physical activity, specific disease risk, and over-all mortality are complex. In general, epidemiological evidence strongly supports the view that physical activity lowers the risk of diseases and reduces the risk of death. The type of activity seems to be of secondary importance, and just the fact of increased energy expenditure is what matters the most. All activity, whether resulting from recreational exercise, housework, or occupational work, reduces risk. Prevention of some diseases, however, requires specific types of exercise. For example, as mentioned earlier, prevention of osteoporosis can only be achieved by weight-bearing activity, and thus swimming, even at a professional level, is not a bone-building exercise (Creighton et al. 2001; Dalkiranis et al. 2006). How the exercise or other activity is performed—over a single longer period or a few shorter bouts—also seems of little importance, although we need to keep in mind that most studies have assessed changes in VO_2max, and not disease risk.

The diverse results of the research can be interpreted in two ways. First, a main problem could be the methodology of epidemiological research. Most studies are retrospective, so levels of physical activity in the past, often over an entire lifetime, are assessed by questionnaires. Not only can such assessments be biased due to problems with remembering activity from the past, but also different types of questionnaires are used by different studies, which make comparison of the results difficult.

But it is possible that the variability in results—showing that basically what is important is merely to be active, and the more activity the better—reflects a real phenomenon. Just as humans are adapted to a wide range of dietary environments, they may be equally adapted to a wide range of physical activity. Studies of contemporary hunting-gathering societies would seem to support this hypothesis. Hunter-gatherers differ, first of all, in their overall levels of expanded energy. Whereas such levels are high for the Ache, among the !Kung they are relatively low, especially for women. Further, a gender division of labor is clearly present in most hunter-gatherer societies. As Frank Marlowe (2007) wrote, "Males and females target different foods and share them." Both men and women may hunt small animals, but big game hunting is almost exclusively done by men, with the exception of the women of the Agta (hunter-gatherers from the Philippines), who hunt with the help of dogs (Estioko-Griffin and Griffin 1981). Gathering is done by both men and women, although men's involvement depends on environmental conditions; men do more gathering in less seasonal, more productive habitats (Marlowe 2007).

Sexual division of labor implies that men and women performed, to some extent, different types of physical activities, partially due to the different tasks associated with getting food but also resulting from the demands of reproduction and child care for women. For example, humans perform remarkably well at endurance running (Bramble and Lieberman

201

2004), but walking and not running was clearly the mode of locomotion preferred by women who were carrying children or were in an advanced state of pregnancy.

Moreover, among hunter-gatherers the duration and types of activities performed show a pattern of seasonal and weekly variation. During some days of the week, people went on physically demanding hunting and gathering trips, but other days were spent in camp—eating, resting, and socializing. Men usually hunted from two to four nonconsecutive days per week, and women gathered every two to three days (Eaton and Eaton III 2003). Inactive days were clearly preferred over days when people had to work, and the preference for inactivity can be interpreted as an obvious adaptation to conserve metabolic energy (Williams and Nesse 1991). Physical activity in evolutionary times was not the preferred choice, but a necessity—there was no way to obtain food other than through being physically active. This human preference for inactivity is perhaps the basis of the difficulty modern humans have in starting an exercise program and following it through for a long period of time.

It is important to realize that our evolutionary ancestors were physically active not because they enjoyed exercise but because their lifestyle necessitated engaging in a lot of energy-burning behaviors. Natural selection promoted strength and endurance, and our physiology evolved to allow our ancestors to perform the many different activities that were required to find and process all kinds of foods, to escape predators, and to take care of children. That means modern humans have inherited a physiology that can handle high levels of diverse physical activities. But this human ability to be physically active did not evolve to prevent the diseases of middle and old age. As discussed previously, natural selection is much more concerned with honing the health and reproductive potential of young individuals than it is with maintaining good health in postreproductive years. The evolutionary approach and knowledge about the lifestyles of our ancestors cannot tell us which activities of what duration are needed to prevent modern diseases.

Although epidemiological and experimental studies provide important information, they are usually concerned only with selected physiological, metabolic, or health aspects of physical activity. Adding an evolutionary approach to the medical and epidemiological framework could create a more comprehensive picture of the relationship between physical activity and health. At the same time, it would be naïve to think that research enriched by an evolutionary point of view would yield precise recommendations about the specific activity patterns needed to preserve good health.

So far we have discussed trade-offs in physiology, trade-offs between reproduction and survival, and trade-offs in the risks of various diseases. But no discussion of human biology would be complete without considering the impact of culture. Humans have invented specific ways of dealing with the severity of biological trade-offs in the form of cultural practices, customs, and beliefs to circumvent, minimize, or neutralize the trade-offs. Ironically, sometimes such practices have made the trade-offs worse. Chapter 10 will take a quick look at selected examples of cultural practices from the point of view of an evolutionary reproductive ecologist.

10

EVOLUTIONARY TRADE-OFFS

AND CULTURE

Cultural beliefs and practices may help people deal with biological trade-offs, but some practices are definitely detrimental to biological fitness. Some aspects of culture relax biological trade-offs to the extent that survival or reproduction is enhanced, but many customs cannot be as easily classified as either enhancing or lowering the biological "quality" of individuals, and these customs' interactions with human biology are complex. Thus, anthropological literature frequently speculates about the purpose served by these cultural beliefs and customs, and often many explanations are provided for the observed phenomena (though I will not attempt to review all anthropological interpretations).

For example, female genital mutilation, a practice common in many societies, endangers women's well-being, health, and even life. Although the reasons for subjecting women to such procedures are complex, some anthropologists have suggested that female genital mutilation is practiced in societies that highly value female chastity and fidelity (Dickemann 1979, 1981), where it is supposed to help ensure a high probability of paternity. However, some societies believe that it is necessary for the women themselves, not just for men concerned with the chastity of their wives: circumcised women are believed to be more likely to observe postpartum taboos—other cultural practices of potential importance for the health of mothers and children alike.

This chapter focuses on several such examples of cultural practices that have an impact on human reproduction, but it is irrelevant here whether the practices are genetically determined or passed on by other modes such as learning or observation. Further, I am not interested here in the frequently discussed issue of gene-culture coevolution, which explains human behavior as resulting from interactions between genetic evolution and cultural evolution. I will also not attempt to analyze whether practices that seem to increase reproductive potential appeared in many societies because of the biological purpose they serve or appeared for a different (cultural) reason and have been maintained ever since solely for cultural advantages. Finally, I will not attempt to understand why practices that are clearly detrimental to reproductive success are so widely popular in many populations or religious groups. These issues are of great interest but are beyond the scope of this book.

Postpartum Sex Taboos

Postpartum sex taboos, common in populations all over the world, seem designed to help women deal with the costs of reproduction. Women are not allowed to get pregnant too soon after having a baby. These taboos extend interbirth intervals, providing the mother with a chance to improve her nutritional condition before the next pregnancy imposes a new energy drain.

Wagogo women, agro-pastoralists from the Dodoma District in Tanzania, breastfeed until a child is between twenty-four and thirty months old (Mabilia 2005, 51). The decision to stop breastfeeding is made by the mother when the child has reached "the right age" or "is able to walk quickly" (Mabilia 2005, 51). With a mother's final child, breastfeeding may last until the child is almost five years old, and mothers nurse, according to one of them, "for her own pleasure and that of the child" (Mabilia 2005, 51).

The Wagogo believe that sexual activity during pregnancy is necessary for the growth of the fetus, and both husband and wife "must look after the pregnancy" (Mabilia 2005, 70). Looking after the pregnancy is different from practices in many Western countries. The woman perceives the period of gestation as a normal part of her life and does not make any special adjustments either to her workload or diet (Mabilia 2005, 69). Instead, the future parents look after the pregnancy by engaging in sexual activity up to the fifth month of gestation. The husband's sperm mixed with the woman's blood is believed to nourish the fetus and help the future baby grow.

After the child is born, however, strict rules prohibit any sexual activity (Mabilia 2005, 81). The mother must not have sexual intercourse as long as she is breastfeeding the child. "A Gogo mother who becomes pregnant during the period dedicated to breastfeeding (worse still if this occurs during the first year of the newborn baby's life) is severely judged by all the other women—'a very ugly story, *ilema ibi sana*: it is the most evident proof of the mother's lack of care towards her newborn baby' " (Mabilia 2005, 82–83). It is believed that not following the rules of sexual abstinence may result in serious problems for the child's health. In fact, any health problems the baby is experiencing, including commonly occurring diarrhea, are likely, according to other women, to result from maternal lack of sexual abstinence.

As soon as the woman discovers that she is pregnant again, she is forced to stop breastfeeding. This custom, of course, makes sense from the physiological point of view. She is prevented from metabolizing for three, which could easily lead to maternal depletion and perhaps lower the energy flow to the developing fetus. "One woman's words describe well the condition of a pregnant mother with a small child to raise: A woman is only one person and she has to carry two weights: one in her stomach and one on her back. A truly onerous and shameful condition, a situation to be avoided. Only one baby at a time should be raised" (Mabilia 2005, 86).

Similar beliefs are held by many societies. The Luo, an agricultural people who subsist on maize, sorghum, and millet, live in Western Kenya, a region with high rates of early childhood mortality (Cosminski 1985). People there have a very good understanding of the energy allocation dilemma. They believe that the milk of a pregnant woman is poisonous to a nursing baby and will cause it to get the illness *ledho,* which has symptoms typical of protein-energy malnutrition. The Luo people are also convinced that nursing while pregnant is dangerous to the nutritional condition of the fetus because the infant is taking milk from the fetus. They also consider the practice harmful to the mother, "making her thin and tired."

In farming communities in central Mexico, the mother must stop lactating as soon as she learns that she is pregnant to avoid damaging the developing fetus (Millard and Graham 1985). In central Java, the Ngaglik people, who live in one of the most densely settled rural areas in the world, have established "strong negative sanctions against sexual intercourse during at least the first year, if not longer, of breastfeeding" (Hull 1985). Just like in many other societies, they believe that intercourse may negatively affect the health of the nursing baby because it changes the quality of milk. There is even a special Javanese term *kesundhulan* that refers to pregnancy during lactation (literally it means the displacement of one child by the next). In Ghana, *kwashiorkor* (meaning "a rejected one") is a disease of the present baby when the next one is born. In the Fiji Islands, a child

nursed by a pregnant mother is believed to develop the life-threatening illness *save,* which makes it lose its appetite and causes its legs and body to
become weak (Katz 1985). The Igorot from the Philippines believe that
after the mother becomes pregnant, her milk turns oily and difficult to digest, so the nursing infant will get diarrhea. Women say that the best litmus
test to detect a new pregnancy is diarrhea in a nursed baby (Raphael and
Davis 1985, 36). "The belief that a pregnant woman should not breastfeed
may serve to protect both mother and child in societies such as Sagada,
where many adults have an 'unbalanced' diet or simply don't get enough
to eat" (Raphael and Davis 1985, 37).

The Usino women from Papua New Guinea think that when semen
comes in contact with milk inside the woman's body, her milk will turn
yellow or black, and it will make the child sick (Conton 1985). If milk is
contaminated by semen, they further believe, the child grows slowly, has
tiny buttocks and chickenlike spindly legs, and is generally poorly developed. In Maisin villages, also located in Papua New Guinea, children
whose parents had sex while the woman was lactating are said to have
large heads and stomachs, and small hips and legs (Tietjen 1985). Among
the Usino, the minimum socially acceptable birth interval is two years; it
was suggested that in the past, during the precontact period, postpartum
sex taboos lasted much longer, for up to four years. For Usino fathers,
who were involved in frequent raids on neighboring villages, large families and too closely spaced children were problematic. Young children
were more vulnerable during raids, and a father involved in fighting did
not have time to plant gardens large enough to support his expanding
family. However, peace treaties with their neighbors in the 1920s and
1930s changed the duration of the Usino postpartum taboos, making abstinence shorter by a year or two (Conton 1985)—politics can ruin our
sex lives but can improve them, too.

In the past, similar taboos against sex during lactation also existed in
European countries. In fifteenth-century Florence, the merchant class
held a belief that an infant might die if breastfed by a pregnant mother
because pregnancy made her milk go bad (Maher 1992). In seventeenth-
century France, the practice of wet-nursing was encouraged by the Christian church to remove the need for postpartum sex taboos (Reynolds
and Tanner 1983, 38). At that time in France, a general belief was held that
sexual intercourse could interfere with the milk production of a nursing
mother. Sending infants to wet nurses, usually to the countryside, freed

mothers from the necessity of breastfeeding and allowed them to resume marital sexual relationships. Even though wet-nursing greatly increased infant mortality, it was still supported by the church, which felt that this was the only way to prevent husbands from having extramarital affairs: "the wife should, if she can, put her children out to nurse, in order to provide for the frailty of her husband by paying the conjugal due, for fear that he may lapse into some sin against conjugal purity" (Reynolds and Tanner 1983, 39).

Getting pregnant while lactating might have been too much even for women with good nutritional status. In many populations, there are additional constraints on energy allocation to reproduction. Catherine Panter-Brick, who worked with the Tamang, an agricultural community in Nepal, observed that women with a high workload could not afford to reduce work intensity when nursing, especially during the harvest season when a woman's work was too essential for the household economy (Panter-Brick 1992). In the small villages of southern Poland, pregnant women were expected to work as hard as other members of their families during the harvest. Among the Igorot people in Sagada villages of the Philippines, a pregnant woman is supposed to carry her usual workload until the time of delivery "even when she felt nauseated, weak or dizzy" (Raphael and Davis 1985, 30). After she gave birth, she was allowed to do nothing but take care of her baby, and her relatives would tend her fields. After two months, she was expected to resume her usual workload (Raphael and Davis 1985, 33).

In Oceania, where postpartum sex taboos are nearly universal (Gussler 1985), people cite several reasons why the taboos should be observed. The most common beliefs are that the husband will lose his strength and health from the lactating woman's polluting influence, the mother herself may become ill, and children's health will also be negatively affected (Barlow 1985).

In many societies, additional complications result from breaking postpartum sex taboos. When the mother gets pregnant again, she faces the problem of feeding her child, who is not allowed breast milk any more. Providing appropriate food, especially during the dry season, adds additional energy-depleting tasks to the woman's already intense daily workload (Mabilia 2005, 86–87).

Lesthaeghe, Meekers, and Kaufmann (1994) claim that the occurrence of postpartum sex taboos differs dramatically between hunter-gatherers and pastoralists versus agriculturalists. Such taboos are practiced in about 23 percent of hunter-gatherer and pastoralist populations, and in up to about 85 percent of agricultural populations.

Postpartum taboos would be expected more in agricultural than in hunting-gathering populations. Among hunter-gatherers, there is a limited need for cultural practices that ensure sufficiently long interbirth intervals. Intense, prolonged lactation with frequent nursing bouts is likely enough to suppress ovarian function, especially for women in relatively poor nutritional condition. In many agricultural societies, maternal nutritional conditions may also be poor (or often even poorer), but there also are substantial seasonal fluctuations in both energy intake and energy expenditure.

In many food-producing populations, the periods with improved nutritional conditions that occur after harvests are long enough for the resumption of ovarian function to take place, despite continuous breastfeeding. A woman could therefore conceive while lactating, with all the negative physiological consequences of "metabolizing for three." Thus, a postpartum taboo removes the danger of a new pregnancy occurring too soon for a nursing mother, because her improved nutritional condition lasts just for a short time.

As Fat as Possible

> She is beautiful to the eyes, oh my lord, and God gave her,
> a breast new and green appearing like two balanced
>   weights . . .
> Gave her a waist lined with stripes,
> Gave her a thigh with stretch marks reaching from her
>   stomach to her knee . . .
>
> —Fragment of a poem recited by Boukia
> at Tchin Tabaraden, Niger, 1990 (Popenoe 2004, 135)

Fatness, stigmatized in many contemporary Western societies, is a symbol of self-worth and sexuality in many nonindustrial populations (Brown 1991). In such populations, due to more traditional diet and activity patterns, many of the health problems associated in industrial populations with fatness are less likely to occur.

Among the Zarma in Niger, married women hold very peculiar contests after the harvest season. The one who becomes the fattest wins, especially if she is able to accumulate rings of fat around her neck (Popenoe 2004, 6). Among the Moors, the Sahara desert pastoralists, a woman is considered beautiful if she is as fat as possible (Popenoe 2004, 1). The Tarahumara of northern Mexico consider fat legs in a woman the most important aspect of ideal beauty, and an attractive woman is called "a beautiful thigh" (Brown 1991). Among the Kipsigis of Kenya, fatter girls generate a higher bride price (Borgerhoff-Mulder 1988). In Jamaica, fatness is a trait believed to be related to happiness, vitality, and health (Sobo 1993, 32). Even in the modern United States, among Puerto Ricans in Philadelphia, obesity is not only socially acceptable, but a married woman's fatness is perceived to be a sign of a good marriage, with a husband capable of taking good care of his wife (Massara 1989). Being at least plump is desirable in 81 percent of societies on which appropriate data exists in the Human Relations Area Files (Brown 1991), a large set of ethnographical and archeological data that has been compiling since 1949.

In some societies, girls are intentionally fed to become fat, often by drinking a lot of milk and observing restrictions on physical activity. Among the Moors, girls are given a fattening diet for several years before puberty, usually from the age of eight, so that they will accumulate fat and achieve a body size admired by all (Popenoe 2004, 1). In other societies, women are fattened for a shorter period, often just before marriage. The Efik of Nigeria put girls in special fattening huts for two years (Malcolm 1925).

The Moors are quite remarkable in that "fattening begins so young, goes for so long, and is in fact the central preoccupation of women's life" (Popenoe 2004, 6). The main diet used for fattening consists of milk and grains, mostly millet and sorghum, so that it is relatively low in fat but high in carbohydrates. Not all girls, especially in the past, were exposed to the fattening procedure. Such a procedure required a lot of food, and girls were not supposed to contribute to any energy-burning agricultural or household activities. Therefore, usually only girls from rich families could afford to get fat. However, being fat was so important that even less-well-off families fattened their daughters despite the high costs, especially in cases when the mother of the girl was fat herself or when the girl's marital prospects were poor—that is, when her father was dead, or the family did not have close ties to the community, or when the girl was not considered attractive (Popenoe 2004, 45).

In the past, a girl was given a milking cow by her father and was supposed to use the milk for her fattening process (Popenoe 2004, 46). Girls were almost never fattened by their mothers but rather by other female relatives, often the paternal aunt. The woman in charge was responsible for making sure that the girl was eating a lot and would use all possible methods, including physical force if needed.

> Little Aichatou was five or six years old when I met her, and was being raised lovingly by her grandmother, Fatima . . . One day . . . I entered Fatima's tent to find a rather different scene, a changed relationship between grandmother and granddaughter. Aichatou sat in a corner, her head hung low, a disproportionately large bowl of porridge before her. Every so often Fatima looked over and yelled, "What are you waiting for?" "Eat!" and "Swallow! . . . Although I never witnessed Fatima employ tactics commonly used against other girls being fattened—throwing household objects at them, pulling their fingers back to make them swallow, or pinching the back of their necks—clearly the emotional timbre of Fatima and Aichatou's relationship had changed drastically. . . . From then on for the next six or seven years, under her grandmother's watchful eyes, Aichatou's main purpose in life would be to fatten her body. (Popenoe 2004, 41)

But why?

"As many Arab women told me, of course fattening has nothing to do with fertility, since they have seen many thin women who have no problem having children. Instead, fattening has to do with making a girl into a sexually mature and desirable adult woman" (Popenoe 2004, 45). Although body fat may indeed have a limited relationship to reproductive potential in adult women (and too much fat may even reduce fecundity, as was discussed earlier), the procedure of fattening the girls may enhance their long-term reproductive performance. Nutritional status in early childhood is closely related to the age at menarche, and well-fed girls, especially when they have low levels of physical activity, grow faster and mature sooner. Early maturation not only extends the duration of the lifetime reproductive period but is also related to higher levels of ovarian hormones and a higher frequency of ovulatory cycles in the years after menarche (Apter and Vihko 1983; Vihko and Apter 1984). This prepuberal fattening may, therefore, increase the girl's lifetime reproductive potential.

Short-time fattening, just before marriage, may also have some functional, reproductive significance. In particular, for women who, in general, have rather poor nutritional status, this intense energy shot may serve to put them in a positive energy balance, increase levels of reproductive hormones, and reduce the waiting time from marriage to first conception. The woman who takes a short time to conceive for the first time and thus proves she is fertile will be considered more valuable as a wife and, therefore, worth investing in by her husband.

A process of fattening may also be a way to show that the girl is physiologically capable of accumulating fat—an important feature in populations with frequently occurring food shortages. Food shortages have been known to occur in 118 surveyed nonindustrial societies (Whiting 1958). About 50 percent of populations experienced at least annual or more frequently occurring problems with food, resulting in high fluctuations of body weight, especially in agricultural societies (Hunter 1967). Some researchers have speculated that, due to a high frequency of food shortages, it is possible that natural selection has favored individuals who are especially good at accumulating fat (Brown 1991). Variation in such an ability may be genetically based, and, therefore, a woman who is able to get fat herself may pass on this valuable skill to her children.

Fatness, somewhat surprisingly, may also be related to psychological well-being. Admittedly, in European and U.S. populations, several studies have suggested an increased risk of depression for obese people (Roberts et al. 2000, 2003; Johnston et al. 2004), but others show that obesity is associated with depression only among the severely obese (Dixon et al. 2003; Onyike et al. 2003; Dong, Sanchez, and Price 2004).

In other populations, the "Jolly Fat" hypothesis seems to hold, which proposes that overweight people of both sexes have a lower risk of depression and a reduced number of depressive symptoms (Crisp and McGuiness 1976). Among the Chinese elderly, both obese men and women are less likely to exhibit depressive symptoms than elderly people of normal weight (Bin Li et al. 2004). In Poland, in postmenopausal women who are not well-educated, there is an inverse relationship between body mass index (BMI) and depressive symptoms—that is, women with a higher BMI are more "jolly" (Jasienska et al. 2005).

Although the exact mechanism is unknown, it is possible that in postmenopausal women estrogen may serve as an intermediary in the relationship between body fat and depression. Postmenopausal women have, on average, much lower levels of circulating estrogens than premenopausal women because the ovaries that are the main source of estrogens essentially stop producing steroid hormones. However, postmenopausal women still can have relatively high levels of estrogens in specific peripheral sites, including the breasts, bones, and the brain (Simpson 2002; Simpson et al. 2002).

Levels of estrogens have a direct positive relationship with the amount of adipose tissue (Kirchengast 1994; Hankinson et al. 1995; Verkasalo et al. 2001). This tissue serves as a source of aromatase—an enzyme that converts androgens, produced by the adrenal gland, to estrogens. So fatter postmenopausal women may have higher levels of estrogens than postmenopausal women with less body fat.

Estrogens are the crucial link between fatness and jolliness because they have the ability to alter the function of the neurotransmitter serotonin (Bethea et al. 2002), which, in turn, plays a major role in depression (Archer 1999). In societies where there is no social stigma against obesity, during the years after menopause women with more body fat may have more estrogens and enjoy a lower rate of depression.

Female Genital Mutilation

Female circumcision, known also as female genital mutilation (FGM), involves procedures that range from removing the clitoris to infibulation—cutting off not only the clitoris, but also the labia minora and at least two-thirds of labia majora. After these parts are removed, the two sides of the vulva are pinned together, using sutures made of silk, catgut, or thorns. In some societies, stitches are not used, but various substances that are supposed to help with healing are applied to the wound, including (in some populations in the Western Sudan) animal excrement (Dorkenoo 1994, 5). A small opening is left for the passage of urine and menstrual blood. Female genital mutilation is usually performed on young girls well before puberty (Mackie 1996). After the procedure, the girls' legs are bound together from hips to ankles, and she is not supposed to move them for up to forty days, allowing scar tissue to develop.

The short-term complications of FGM include pain during "surgery," hemorrhage from the rupturing of the blood vessels of the clitoris, and postoperative shock. Sometimes there is also damage to other organs, including the urethra, anal sphincter, and vaginal walls (Dorkenoo 1994, 13–14). Cutting is often performed without sterilized instruments, and girls may contract tetanus, septicemia, or, more recently, human immunodeficiency virus (HIV) or hepatitis B infections. Reopening the vulva after marriage often causes further damage (Dorkenoo 1994, 14). Long-term problems associated with FGM include a lack of sexual sensation, painful menstruation and intercourse, and often excruciating pain accompanying scar tearing during childbirths, prolonged and obstructed labors, and hemorrhaging (Mackie 1996).

Female genital mutilation is practiced in at least twenty-eight African countries. In some of these countries, almost all women go through FGM; in others, the practice is very rare. Over 100 million girls and women are genitally mutilated in Africa alone, where 2 million girls per year—6,000 each day—are subjected to FGM (Dorkenoo 1994, 31). In Somalia and Djibouti, 98 percent of women undergo FGM, in Kenya and Liberia 50 percent, and in Uganda no more than 5 percent (Dorkenoo 1994, 88). Statistics are difficult to obtain, but it is estimated that about 15 percent of all circumcised women die of bleeding or infection (Dorkenoo 1994, 15).

Although FGM is most common in Africa, the procedure was also practiced in other countries. For example, the Skoptozy, a Christian sect in Russia, practiced FGM to ensure virginity (Dorkenoo 1994, 29). In twelfth-century Europe, chastity belts, introduced by the crusaders, allegedly served a similar purpose and were used by husbands to ensure that wives remained faithful while they were away fighting for Christianity. In nineteenth-century London, Dr. Isaac Balzer Brown recommended removing the clitoris as a cure for numerous diseases ranging from insomnia to "unhappy marriage" (Dorkenoo 1994, 30). Today, private clinics in London, while not offering FGM, advertise restoration of virginity for the price of about US$3,000. A woman's virginity is so important for Japanese men that women there undergo hymen repair surgeries at the cost of over US$6,000 (Dorkenoo 1994, 31). In Macedonia, many women decide to have their hymen restored, even though this procedure is both forbidden by law and very expensive: a single surgery costs US$550—the equivalent of two months' average salary.

Why is FGM practiced in so many cultures? In most societies, FGM seems to be a custom invented to prevent women from experiencing sexual desire. It is believed that women who lack the ability to experience sexual pleasure are more faithful to their husbands. Of such customs, FMG is perhaps the most drastic, but others include obsession with virginity, claustration of women (as in harems), or foot binding (practiced in China until the beginning of the twentieth century; Mackie 1996), all of which are aimed at limiting a woman's mobility and thus ensuring the husband's unquestioned paternity (Dickemann 1979, 1981).

People from societies that practice FGM often provide explanations that have nothing to do with paternity certainty. The Mossi of Burkina Faso and the Bambara and Dogon in Mali believe that FGM protects the baby during parturition (Dorkenoo 1994, 34). Because the child will die if its head touches the clitoris, the removal of this dangerous organ is a protective measure to preserve the life of the baby. The Bambara extend the danger of having a clitoris even farther: the poison secreted by the clitoris is so strong that it could kill a man during intercourse.

Some societies, like the Yoruba in West Africa, make the connection between female circumcision and the health of the child. They believe that circumcised women are more likely to follow postpartum sex taboos. It is hard to determine whether such drastic measures are indeed necessary to prevent women from having sex while nursing, but, as discussed earlier, long interbirth intervals are clearly important for ensuring the good nutritional status of both the child and its mother. This reasoning may be an additional evolutionary raison d'être for the rise and continuing presence of these practices.

Of course, the fact that a cultural custom can be explained in evolutionary terms does not make it ethically right or acceptable. Regardless of whether FGM is beneficial for the fitness of men (we do not know if it really is) or even for the fitness of women (in some societies, women who undergo FGM have more children than those who do not; Strassmann and Mace 2008), there is no excuse for subjecting millions of women worldwide to lifelong pain and serious health problems.

Body Covering and Vitamin D

In many countries, usually in accordance with religious traditions, women wear concealing clothing. In Morocco, for example, most women leave their houses either completely veiled or wearing a scarf to cover their heads along with a *Djellaba*—a long, loose-fitting, hooded, long-sleeved garment that is worn to cover more causal clothes underneath (Allali et al. 2006). Investigators have shown that the risk of osteoporosis in postmenopausal years is higher in Moroccan women who wear traditional dress than in women who choose other types of outfits. The reasons for the higher risk can only be surmised, but it is likely that poor production of vitamin D is responsible.

Among other factors, the health of bones both during development and in later life depends on a sufficient supply of vitamin D. The importance

215

of vitamin D in bone mineralization depends in turn on its critical role in the absorption of calcium and phosphorus (Hatun et al. 2005). Vitamin D deficiency impairs the intestinal absorption of calcium, which causes poor bone mineralization. Vitamin D deficiency is thus responsible for bone resorption and osteoporosis; severe deficiency often leads to mineralization defects: rickets in children and osteomalacia in adults (Lips 2007). Deficiency may also increase the risk of many other diseases such as tuberculosis, rheumatoid arthritis, multiple sclerosis, inflammatory bowel diseases, hypertension, and several types of cancer (Zittermann 2003).

Vitamin D in the body comes from two sources: dietary intake of vitamin D and synthesis in skin exposed to sunlight, in particular to ultraviolet B (UV-B) radiation. Dietary intake, unfortunately, is unlikely to provide sufficient amounts of vitamin D because most consumed foods have a very low concentration of this vitamin. Therefore, synthesis by the body is necessary to cover most of the requirements. Cutaneous production depends on many factors such as age, season of the year, time of the day, latitude, and use of sun screens (Saadi et al. 2006). Sometimes, however, culture interferes with vitamin D production, leading to deficiency and, consequently, to health problems.

In Turkish women who cover their entire body, the concentration of vitamin D is lower than in those who have only their faces and hands exposed (Hatun et al. 2005), but no such difference is found among similarly clothed Emirati women (Saadi et al. 2006). The investigators speculated that Emirati women, due to their darker skin pigmentation, need higher doses of UV-B for vitamin D synthesis: the small part of skin exposed is insufficient to achieve a beneficial vitamin D effect. In Beirut, severe vitamin D deficiency was almost three times more common among women who dressed in traditional, concealing clothes than in those who preferred or were socially permitted to wear less covering outfits (Gannage-Yared, Chemali, and Yaacoub 2000). In the group wearing traditional clothes, 84 percent of women were vitamin D deficient.

In Israel, vitamin D status assessed in women two to three days after childbirth was much poorer among Orthodox Jewish mothers than among non-Orthodox Jewish mothers who wear less concealing clothes (Mukamel et al. 2001). Among the non-Orthodox mothers, vitamin D insufficiency was more common in the winter than in the summer; in Orthodox mothers, no seasonal variation in their poor vitamin D status was found. This suggests that, indeed, the type of clothing worn by women significantly limits the skin's exposure to sunlight even during the bright Israeli summers, consequently curtailing synthesis of vitamin D.

In Turkey, about 80 percent of pregnant women are vitamin D deficient (Hatun et al. 2005). Poor vitamin D status in mothers, unfortunately, leads to equally poor vitamin status in their children. In India, vitamin D deficiency in mothers has been linked to a low concentration of this vitamin in the umbilical cord, suggesting that the amount of vitamin D transferred to the fetus was equally low (Sachan et al. 2005). The infants of Australian mothers who had vitamin D deficiency in pregnancy were also deficient, especially if they were breastfed and not bottle-fed (Thompson et al. 2004).

Insufficiency of vitamin D in pregnant women may also lead to an inadequate calcium supply for the fetus. Mothers are equipped with physiological mechanisms to meet fetal needs for calcium. A major maternal physiological adaptation is a substantial, almost doubled increase in intestinal absorption of calcium occurring in early pregnancy (Kovacs 2005). The fetus does not need much calcium so early in pregnancy, thus an increase in intestinal absorption may allow the maternal skeleton to store more calcium in expectation of increased fetal demands later in pregnancy. But not much will happen without vitamin D, which is needed to mediate the increase in intestinal absorption. When the vitamin D concentration is low, mothers may not be able to store calcium even when having a calcium-rich diet.

In cultures where women do not expose much skin to sunlight, girls may be at particularly high risk. They are not only born with vitamin D deficiency due to the concealing clothing their mothers wore during pregnancy, but they start to wear equally concealing clothing themselves, which makes their poor vitamin D status even worse. In Turkey, about 50 percent of adolescent girls wear clothes that cover them completely, and they do not spend much time outdoors (Hatun et al. 2005). In Iran, the vitamin D concentration in boys aged fourteen to eighteen is twice as high as in girls of the same age (Moussavi et al. 2005). Only 2 percent of boys compared with 20 percent of girls were classified as being vitamin D deficient. Of course, boys are not required to entirely cover their bodies.

Vitamin D may also be important for the immune system, and thus its insufficiency may lead to infections. For example, the incidence of tuberculosis is high in Asian immigrants in the United Kingdom, especially in the first few years after migration (Zittermann 2003). It has been suggested that these people became infected in their countries of origin, but a constant exposure to sunlight and resulting sufficient vitamin D synthesis did not allow the development of the disease. After they moved to the United Kingdom, the reduced UV-B exposure led to impaired synthesis of vitamin D, and its levels became insufficient for the proper functioning of the immune system.

Finally, socioeconomic factors add an additional difficulty to finding a solution to vitamin D deficiency in women. The populations where women most strictly follow religious rules of concealing their body are also the poorest. These women and children cannot afford to purchase vitamin D supplements (Allali et al. 2006), and often they are not even aware that such supplements are needed.

As in many other aspects of women's lives, trade-offs are involved in this as well. Protecting the body from sun exposure may have some benefits for reproduction. Exposure to sunlight causes photolysis, the breakdown of folate. Folate is essential to the human body's function because it takes part in DNA synthesis, repair, and methylation. For this reason, folate is also crucial during pregnancy, and its deficiency is a risk factor for birth defects, especially neural tube defects. Dark skin pigmentation during human evolution may have developed in regions with high sunlight exposure as protection against folate photolysis (Jablonski and Chaplin 2010). It would be interesting to know whether women who wear concealing clothing are better protected from folate deficiency.

Cultural practices affect the lifestyles of people and contribute, together with the environment in which people live, to the variation in lifestyles among populations. Some of these cultural practices related to women's reproduction may have originated because they helped relax evolutionarily imposed biological trade-offs. Closely spaced pregnancies or metabolizing for three (providing for a nursing child, a developing fetus, and the mother's own metabolism) is too energetically stressful for many women. Postpartum sex taboos that extend interbirth intervals are culturally created ways of dealing with this energetic stress in populations where environmental conditions no longer effectively suppress ovarian function during breastfeeding. It can be expected that postpartum sex taboos, if observed, are beneficial for the health of women and their children.

Many other cultural customs, such as wearing clothing that is too concealing, may be detrimental to health of women and children. In some cases, relatively simple public health interventions, without any attempts to change the culture, may be helpful. Women who must wear concealing clothing in public should be encouraged to expose their skin to sunlight in the privacy of their own yards, especially during pregnancy. This behavioral change is virtually cost free and could be very effective, although we should learn more about the folate levels in these women before we can safely suggest it. But not all reproductively imposed costs and trade-offs are so easy to deal with.

The majority of human adaptations evolved a long time ago, when our ancestors lived as hunter-gatherers. With major changes in lifestyle that began with the rise of agriculture, these old adaptations have often become no longer effective. For example, long breastfeeding together with the relatively poor energetic status of mothers suppressed the ability of our evolutionary ancestors to conceive and allowed for long interbirth intervals. With agriculture, weaning food for babies became easily available, thus reducing the need for prolonged breastfeeding. Shorter breastfeeding leads to shorter interbirth intervals, and, in consequence, much worse health for mothers and babies. We have looked at cultural customs and beliefs that, in some populations, interfere—usually unintentionally—with reproductive biology and health. In modern times, many governmental or private organizations also have attempted to interfere with human biology in order to improve the health of women and children. The outcomes of these interventions are not always as expected. In Chapter 11, I will discuss the difficulties encountered by public health and philanthropic efforts that have attempted to improve some aspects of women's and children's health.

11

Gardeners know well that yellow roses are more resistant to frost but at the same time are more susceptible to fungal diseases. Awareness of such trade-offs is enough for a gardener to have a beautiful rose garden, and she or he does not need to know why yellow roses cannot have resistance to both problems. In areas with little danger from frost, a problem-free rose garden can be designed by planting roses of pink and red colors. From health professionals working in all areas related to human health we should, however, expect more than from amateur gardeners. Health practitioners should be aware that in human biology trade-offs exist, and they should know why they exist and understand why this knowledge is crucial for their profession.

A lack of understanding of the evolutionary logic that underlies some trade-offs can be quite detrimental. Public health interventions, especially in the area concerned with improving the health of mothers and children, can bring unexpected outcomes, sometimes simply not helping, sometimes even changing the situation of the targeted population for the worse. The identification of the problem, good intentions, and plentiful financial resources are all important, but often are not enough.

Sometimes the lack of relatively simple knowledge about the lifestyle of the people who are in need of help has been responsible for failures of philanthropic or governmental actions. For example, iron-deficiency anemia is a serious problem for many populations around the world, especially in developing countries. This deficiency has many negative consequences for the mental and physical development of children and can result in adverse outcomes in pregnancy as well as decreased work capacity and productivity. Lack of another nutrient, folate, is associated with a higher risk of neural tube defects in developing fetuses, leading to mental retardation. In the early 1960s, several Central American countries, includ-

220

ing Guatemala, began to fortify wheat flour with iron; in 2002, folic acid and a few other nutrients were added (Imhoff-Kunsch et al. 2007). In Guatemala, where many people are iron and folate deficient, programs of flour fortification ought to have been valuable, but most nutrient-deficient people in Guatemala live in poor, rural areas where the main staple is corn, and wheat flour is consumed only in low quantities. Thus, the programs of flour fortification implemented forty years ago had limited impact on improving the nutritional status of the people who needed it the most.

The underlying reasons for the failure of the flour fortification program in Guatemala are easy to understand, but other programs designed to improve health conditions have had outcomes that were not only unexpected but also difficult to explain without understanding the principles of evolutionary biology.

Supplementation Backfires: More Babies instead of Healthier Babies

Many international agencies are concerned with improving children's nutrition, especially in developing countries. The World Bank and the United Nations Children's Fund (UNICEF), for example, see such improvements as an important tactic for changing the quality of human life and promoting economic growth (Martorell 1996). Good nutrition should be available to children as early as possible, preferably during fetal development. Often this is not possible. In the rural Gambia, women give birth to children with low body weights, which would not seem to be surprising given the impaired nutritional status of the country's poor farming communities. The rural women have rather high levels of physical activity and low caloric intake (Lawrence and Whitehead 1988; Singh et al. 1989), especially during periods of seasonal food shortages when they consume approximately 1,000 kilocalories per day. They often cannot increase their dietary intake during pregnancy and lactation, when their energetic needs greatly increase. What if these women could get more food during pregnancy? It seemed obvious that if they had such an opportunity the birth weight of their babies would increase.

In the 1960s, the Medical Research Council of Britain and the Dunn Nutrition Laboratory of Cambridge University established in the Gambia a long-term program to monitor the nutritional status of mothers and children and to design interventions to lower the mortality rates of children (Prentice et al. 1981). Low birth weight carries with it a high risk of childhood mortality, especially during the first year of life; therefore, improving birth outcomes became one of the main targets of the program. Pregnant women received nutritional supplements, which added substantial energy, protein, and nutrients to their daily diet (Prentice et al. 1987). Their energy intake during pregnancy increased daily by approximately 725 kilocalories, which was a substantial increase in energy intake; during the most expensive third trimester of pregnancy, maternal energetic requirements increase, on average, by 470 kilocalories per day. A significant improvement in birth weight was thus expected. However, although birth weight did indeed increase after the supplemented pregnancies, the increase was surprisingly low. On average, the supplemented mothers had babies weighing just 50 grams more.

During the lactation period, the additional food that the mother received had little impact as well. Some women who had been supplemented during pregnancy received additional food during subsequent breastfeeding, and others were supplemented only during lactation. It was expected that supplementation would improve milk quantity and milk composition, but neither changed significantly. Their milk output remained exactly the same; while their milk had a little more protein, its energy content was the same as that of the unsupplemented mothers. Peter Ellison explains this finding by noting that milk production is well buffered against variations in maternal nutrition (Ellison 2003b, 92). Mothers maintain high-quality milk even when they are undernourished, getting the necessary energy from fat that was accumulated during pregnancy and reducing their own metabolic requirements.

If the energy from the supplementation did not go to the baby in utero, and it was not used to improve the energy content of the milk, what happened to it? It appeared that supplemented women had shorter interbirth intervals after a calorie-supplemented pregnancy (Figure 11.1) (Lunn, Austin, and Whitehead 1984). This suggests that the energy from the supplement was used by the mother to improve her own nutritional condition, and only a small fraction was allocated to improve the condition of her current child. This observation perfectly illustrates one of the basic trade-offs suggested by life-history theory: between current and future reproduction. When an undernourished mother receives additional resources, most of it is allocated to support her own physiological and metabolic needs. A woman with an improved nutritional condition can invest more in future reproduction. And this is exactly what was observed in the Gambia—supplemented mothers were able to conceive again sooner.

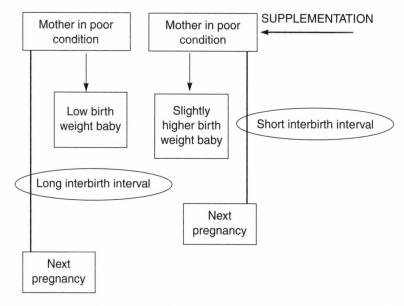

Figure 11.1. Supplementation of women who have poor nutritional status during pregnancy has a limited impact on improving the birth weight of the babies. Instead, it reduces the interbirth interval to the next conception.

Thus, if an organism would follow this strategy during all reproductive years, this would clearly result in a greater number of children but not necessarily in healthier children.

The physiological link between supplementation and the resumption of postpartum fertility was suggested by the measurements of prolactin, a hormone that stimulates the production of milk. In the Gambia, the prolactin levels for up to 80 weeks after delivery differed significantly between the supplemented and unsupplemented mothers. Mothers who did not receive any supplements had the highest levels of prolactin, mothers supplemented during both pregnancy and lactation had the lowest levels, and those supplemented only during lactation had intermediate levels. Involvement in production of milk is not the only role of prolactin, as this hormone also can suppress ovarian function (McNeilly, Tay, and Glasier 1994; Bribiescas and Ellison 2008). In women with good (or merely improved) nutritional condition, prolactin levels drop more quickly; despite continuous nursing, the levels do not rise enough to maintain postpartum infertility. Thus, supplementation leads to lower levels of prolactin and a more rapid resumption of cycles.

Calcium Supplementation Makes Bones Stronger?

Gambian infants not only had low birth weight, but were also shorter and had low mineral bone density, which suggested that perhaps their mothers did not supply them in utero with enough calcium (Jarjou et al. 2006). The recommended intake of calcium for pregnant and lactating women is about 1,200 to 1,500 milligrams per day and, in comparison to these values, pregnant Gambian women had dramatically low dietary intake of calcium, at about 300 to 400 milligrams per day. A fetus requires, on average, 200 milligrams of calcium per day, and during the period of most rapid growth up to 330 milligrams of calcium each day. Because the mother also has her own physiological requirements for calcium, the daily intake of mothers in the Gambia seemed much too low to satisfy both requirements. Not surprisingly, their breast milk had a low calcium content.

Supplementation with dietary calcium seemed like a simple solution. Half out of 125 Gambian women participating in the program received 1,500 milligrams of calcium per day, beginning from the twentieth week of gestation until delivery, and the remaining half received a placebo (Jarjou et al. 2006). It was expected that babies of the calcium-supplemented mothers would be born with stronger bones and greater body length, since calcium is necessary for skeletal growth. It also seemed obvious that the breast milk of the supplemented women would have higher content of calcium. None of this happened. The calcium content of milk measured at the second, thirteenth and fifty-second week of breastfeeding were practically identical between supplemented and unsupplemented mothers. The mineral status of bones was assessed when infants were two weeks old. Infants of mothers who received calcium, and of mothers who did not, had exactly the same bone mineral content, bone width, and bone mineral density. Birth weight, birth length, and head circumference of both groups of babies were also the same, and they gained weight and height at basically the same rate until the fifty-second week after birth.

The idea of maternal-fetal conflict resulting from not quite the same "genetic" interests of the mother and the developing fetus, has been used to explain the curious patterns of calcium metabolism in pregnancy (Haig 2004). This theory is based on the premise that the maternally-derived half of the fetal genome (i.e., that half which was inherited from the mother via her egg cell) and the paternally-derived half of genome (i.e., that inherited from the father) have different evolutionary interests. The maternally-derived part of fetal genome acts, most of all, for the interest of the mother, maximizing her lifetime reproductive success, even when it means worsening the prospects of the current offspring. When the mother is calcium-depleted during pregnancy, the maternally derived alleles of the fetus favor a reduction in calcium flux across the placenta. It is not in their interest to allow further deterioration of health of the mother who, after all, carries copies of these very alleles. Therefore, more calcium is retained by the maternal skeleton, and less is available for the use by the developing skeleton of the fetus. The alleles inherited from the father have an entirely different view of this issue: it is in their interest to affect the metabolism of the fetus in such a way that the flux toward the fetus is maximized.

Because calcium supplementation seems not to improve bone health of infants, does it mean that programs designed to supply calcium to pregnant women in populations with very low calcium intake are not needed? Long-term, prospective studies are necessary before this question can be answered. As discussed before, it is possible that low protein intake interfered with calcium absorption in Gambian women, thus supplementation with both calcium and protein may bring different results. In addition, even if there are no benefits to infants, perhaps the mothers profit from such supplementation? A short-term project of calcium supplementation of pregnant Mexican women suggested that bone turnover in the mother can be suppressed by supplementary calcium intake (Kovacs 2005). Bone mineral content of supplemented mothers should be assessed after several calcium-supplemented pregnancies and their risk of osteoporosis evaluated in post reproductive years. So far it seems that supplementing with either energy or calcium during pregnancy or lactation does not bring expected health benefits either to children or to their mothers, at least in the short term.

Workload Reduction Produces More Babies

A pregnant woman may get some extra energy for the developing fetus either by increasing her intake or by decreasing expenditure. If providing additional dietary energy to women does not help their infants, perhaps reducing maternal physical activity would be more beneficial? In many populations, reduction in energy expenditure is an unlikely option, because women must continue their usual work during pregnancy and while nursing (Raphael and Davis 1985, 30; Institute of Medicine 1992), at least during some seasons (Panter-Brick 1993). In other populations, technological improvements help women reduce time and energy devoted to work activities. In an agro-pastoralist community of Southern Ethiopia, the workload of many women became quite dramatically reduced (Gibson and Mace 2006). In this poor community women used to carry water to their houses over great distances, especially during the dry seasons when they often walked for about 30 kilometers with clay pots full of water. Between 1996 and 2000, in some villages tap stands were installed, changing the time spent carrying water from three hours to just fifteen minutes. Mhairi Gibson and Ruth Mace wanted to find out if easier access to water and the subsequent substantial reduction in women's energy expenditure had any impact on demographic variables.

At first, it seemed that access to water had positive effects, as important changes occurred in the survival of the youngest children. With access to water taps, the monthly risk of child death decreased by 50 percent during the first thirty-nine months of the child's life. Unfortunately, it turned out that children under fifteen years of age became more at risk of being malnourished. I will talk about these effects later on. As for the women, surprisingly, the dramatic reduction in workload did not seem to have any effect on their nutritional condition, as indicated by the lack of changes in their BMI. But perhaps the most striking outcome of water tap access was a remarkable change in the birth rate. Women with easy access to water were more than three times as likely to give birth in a given month than women who had to use up a lot of energy getting water the old way.

As already discussed, energy expenditure resulting from physical work has a suppressive effect on ovarian function. Both estradiol and progesterone levels are reduced and, in consequence, the probability of conception is much lower. When, thanks to easy access to water, Ethiopian women expended much less energy, ovarian production of steroid hormones became, most likely, more intense and an ability to conceive much higher. In this population, before technological improvements were introduced, heavy workloads consistently triggered an effective mechanism allowing women to sustain longer pregnancy-free periods. Without intense work women had reserves of metabolic energy which were promptly allocated into producing new children.

Short inter-birth intervals create several negative circumstances for children. In urban Brazil, children born following intervals shorter than twenty-four months had lower birth weight, a higher rate of postneonatal mortality and poor nutritional status at the age of nineteen months (Huttly et al. 1992). In Ethiopian villages, ever since water taps were installed, the mortality of young children has decreased, but the nutritional status of all children declined. According to the authors, it is likely that the decrease in mortality resulted mostly from improved survival of babies who had low birth weights. Babies with low birth weights had the lowest chance of survival before water tap installment when their mothers had to devote time and energy in collecting water. Mothers released from the burden of bringing water home had more time for childcare, which was especially needed by babies in poorer condition. As more babies survived, however, competition between them over resources increased. The number of children in the family was growing, but food to feed them was not. This would explain a decline in the nutritional status of children in water-tap villages.

In general, these results confirm a fundamental biological principle: if the mother receives additional energy, either by increased intake, or decreased expenditure, she is unlikely to use this surplus to improve the condition of the child she is currently carrying in the womb or nursing. Instead she allocates these resources for the faster production of the next progeny. If children survive and reproduce, this strategy would benefit maternal evolutionary fitness, but may not be desirable from a demographic or policy perspective.

Good Childhood Nutrition Enhances Fertility

Improved nutrition in utero seems to have a very limited impact on improving the condition of the newborn. These findings, however, should not lead to the conclusion that the child's organism does not notice such an improvement. On the contrary. Improved early nutritional conditions may have long-term consequences. Studies on supplementation in pregnancy and during early childhood which have been conducted since 1969 in rural Guatemala have yielded many interesting discoveries. Pregnant women and children younger than three years old were randomly assigned to receive one of two types of supplements: Atole, a high-energy, high-protein supplement, or Fresco with low energy and no protein (Ramakrishnan, Barnhart et al. 1999). Both supplements had the same amount of vitamins and minerals, and were given to people twice per day.

About twenty-three years later, women who received one of these supplements either (indirectly) as fetuses or as young girls, began their

227

own reproductive lives. An amazing set of data has been gathered over the years. Rate of childhood growth, age at menarche, date of first intercourse, pregnancy and birth are known for most of the 954 women. Women who received the high-energy, high-protein Atole had a shorter median interval from menarche to first intercourse, and from first intercourse to first birth. Median age at first birth occurred 1.17 years earlier in the Atole-supplemented than in the Fresco-supplemented women. Age at first birth was also earlier for women who showed better growth during their early childhood.

Evidence from two previously described studies sheds some light on the mechanisms which may be behind the relationship between good energetic condition at the beginning of life and subsequent fertility. First, our own study on Polish women showed that ovarian sensitivity in adulthood is related to a woman's size at birth (Jasienska, Thune, and Ellison 2006) and, therefore, most likely depends on in utero nutrition. A study of Bangladeshi migrants to the United Kingdom documented higher levels of ovarian function in women who, due to lower rate of infections and lower "maintenance" costs of their immune system, experienced better energetic conditions in early childhood (Núñez-de la Mora et al. 2007). Good fetal and childhood conditions seem to lead to higher levels of ovarian function, and, therefore, an increased ability to conceive.

Stunting (that is growth that is too low for a particular age in comparison with healthy reference populations) during early childhood seems to delay menarche in girls. It is often a consequence of poor childhood nutrition. A prospective study (i.e., a longitudinal study involving repeated observations or measurements of the same subjects over long periods of time) from Guatemala shows that stunted girls mature about seven months later than other girls (Khan et al. 1996). Later menarche is, in turn, associated with a lower frequency of ovulatory cycles and lower levels of ovarian hormones (Apter and Vihko 1983; Vihko and Apter 1984). Women undernourished as children may thus have a delayed first pregnancy and perhaps a reduced ability to conceive throughout their lives.

In the Guatemalan study described earlier, two other factors interacted with the effect of supplementation with either Atole or Fresco. Women whose fathers had some education (usually just an ability to read) had, on average, the interval from menarche to first intercourse longer by two years compared to daughters of illiterate fathers. Time from menarche to first birth, and age at first birth were also delayed by about two years for these women. A women's own education was even more important, since completion of primary school increased the time from menarche to first intercourse by about six years and delayed the median age of first birth by more than four years.

Good nutrition seems especially important for young children. An inadequate intake of energy and nutrients during this period of life is associated with delayed cognitive development and poor performance at school (Grantham-McGregor et al. 1997), with possibly long-term adverse outcomes in many aspects of adult life. The follow-up of the Guatemalan study evaluated the effects of early supplementation on educational achievements of young women (Li et al. 2003). Women who received supplementation with either Atole or Fresco in early life took tests on knowledge, numeracy and reading when they were between 22 to 29 years old. Women who received Atole had better educational achievement than women who received the poorer supplement. However, the fact of having completed primary school had a much stronger impact on their educational achievement than the supplementation. Women who had completed primary school performed much better on tests than those who did not, irrespective of which supplement they had received as children. Among women who completed primary school, Atole was related to much better test performance that Fresco, but this richer supplement did nothing for the women's intellectual ability at adulthood if they did not finish school!

Paradoxically, nutritional supplementation of children may lead to several long-term problems. As discussed earlier, low birth weight positively correlates with increased risk of several diseases in adulthood. But the highest risk is present in people who were born with low weight, but experienced a rapid increase in weight or height during childhood. In Helsinki, the highest death rate from coronary heart disease was observed for men who were born as thin babies, but gained weight at a fast rate during early childhood (Eriksson et al. 1999). Therefore, supplementation of children born with low weight, especially when leading to fast catch-up growth during childhood, may mean that these people in their subsequent years may develop cardiovascular disease or diabetes. It is likely, however, that the risk of these diseases will increase only in people who, not only experienced improvement in nutritional condition during childhood, but who also as adults have an excess of energy. In the rural Gambia, for example, people did not show changes in glucose levels, insulin and lipid metabolism after prenatal exposure to malnutrition, even when they had had very low birth weight (Moore et al. 2001). Those people, however, remained in relatively poor nutritional condition during their entire lives.

What conclusions can be drawn from the studies in Guatemala and Ethiopia? Nutritional supplementation of women and children leads to increased fertility and higher birth rate. This is clearly not a desirable outcome from the perspective of public health, especially for poor populations where birth rates are already very high. On the other hand, lack of supplementation of undernourished people leads to many health problems, especially in children and is simply not ethical. Clearly, programs supplementing energy or nutrients are a very important way of improving the quality of life, but they should always be complemented by giving people an opportunity for education and by propagating methods of family planning. The same is true for programs introducing technology which reduce the burden of workloads for women. Without contraception, birth rates are expected to increase when less energy is allocated to physical activity. Better knowledge of evolutionary biology and life history theory would allow us to predict some of these undesirable outcomes of philanthropic interventions.

In developing countries, major health problems are still related to the lack of resources and absence of thoughtful philanthropic actions. In developed countries, disease prevention should lead to understanding the responses of the body to an excess of resources. In both cases, we need an approach that is based not only on physiological and medical knowledge, but also on the principles of evolutionary biology, which can predict the responses of the human organism to lifestyle and environment. Breast cancer is a disease tightly linked to reproduction and overabundance of energy. There are no known prevention tactics that would give women full assurance that they would not become sick from this disease, but certain lifestyle changes could significantly lower the chance that breast cancer will develop.

In the twenty-first century, advocating lifestyle modifications as the main strategy for disease prevention may seem like an old-fashioned approach. After all, have we not entered the era of molecular medicine, when most diseases will be, if not now then in a very near future, cured or even prevented thanks to various kinds of manipulations of our genes? Unfortunately, there is reason to believe that this optimistic scenario is not very realistic.

"Traditional" prevention that attempts to preserve health through changes in lifestyle, mostly in diet and physical activity, is also not free of its own problems. Natural selection, which promotes reproduction and not health in old age, is partially responsible for the lack of a perfect way to maintain good health. In Chapter 12, I will argue that the gene-based approach to disease prevention cannot replace the lifestyle-based approach.

12

Simple Genetic Diseases and Lifestyle Treatments

Most health problems in developing countries still result from inadequate nutrition and excessive work, whereas overabundant energy consumption and lack of physical activity are considered the most health-threatening factors in economically developed countries. Some people believe that the most health problems plaguing economically developed nations will soon disappear or will become much less prevalent thanks to genome sequencing, cloning, and genetic engineering.

Genome sequencing and many other recent advancements in genetics and molecular biology are clearly important for both medical research and medical practice (Chiche, Cariou, and Mira 2002). No one can deny that the treatment of diseases, at least in rich nations, has entered into a new era, but it is too easy to get swept up in the enthusiasm over modern molecular technologies, especially the hopes for what they can do for human health. I emphasize throughout this book that knowledge of evolutionary biology and human evolutionary history is important for understanding many aspects of health and disease. Knowledge acquired from genome sequencing cannot replace the value derived from an evolutionary approach.

Few diseases are determined by simple genetic changes. Phenylketonuria, which is caused by a recessive mutation (i.e., an individual must have both copies of the recessive allele, one from the mother and one from the father, for a recessive mutation to manifest itself), is characterized by a deficiency in the enzyme phenylalanine hydroxylase (Wappner et al. 1999). In healthy individuals, this enzyme metabolizes the amino acid phenylalanine to tyrosine; lack of this enzyme leads to the accumulation of phenylalanine. People with untreated phenylketonuria have im-

232

paired brain development, which usually ends in serious mental retardation. The result of another recessive mutation, Wilson's disease occurs when people cannot detoxify the copper that accumulates in tissues, which eventually leads to liver problems and degeneration of the nervous system (Ferenci 2004).

Both of these diseases are genetic, both are very serious, yet appropriate interventions allow people who have the affected genotypes to lead normal, healthy lives. People with phenylketonuria require a diet free of phenylalanine and supplemented with tyrosine, beginning with substituting maternal milk with special formula. To avoid the symptoms of Wilson's disease, affected people need to stay away from mushrooms, nuts, chocolate, and a few other foods high in copper, and to take pills that help remove copper from their bodies. In countries with good health care, such relatively simple dietary interventions and treatments guarantee a basically symptom-free life.

Can Disease Be Prevented by Replacing Faulty Genes?

The Human Genome Project has promised a different approach to dealing with diseases. To put it in a very simplified way, after a mutation responsible for a disease has been identified, a corrected, "healthy" sequence of the DNA could be made in the laboratory. After the new sequence is injected, the individual's genetic machinery starts to produce the correct protein, thus fixing metabolic problems and curing diseases.

This sounds like a dream come true, but many have criticized the huge financial costs of the genome project. New technological discoveries have substantially reduced the cost of genome sequencing (though we would do well to remember that in many countries even relatively inexpensive medical tests and procedures are beyond the financial means of a significant portion of their populations). Even with lower costs, serious, fundamental questions remain as to whether the Human Genome Project can significantly improve the health of people at the population level. The Human Cancer Genome Project recently has proposed sequencing cancer genomes; its promises of personalized cures for about fifty different cancers have been criticized (Miklos 2005).

An additional complication is the faulty assumption that, when it comes to chronic diseases that have a genetic background, the development of all cases follows the same molecular pathway (Kaput 2004). This is not the case. For example, cancers of the same organ that have a similar

233

morphology and histology may, in fact, be caused by different molecular processes, with different genes involved in different people.

Richard Lewontin (1992, 2000) points to several major problems with the promising scenarios proposed by the genome projects. For example, even among diseases with a clear genetic background various changes in the DNA sequence may cause the same phenotypic effect. A *phenotype* is any observable characteristic of the organism that results from the interplay of genetic information in DNA (genotype) and the influence of environment. Thalassemia is a blood disease in which the abnormal production of the hemoglobin molecule, a protein responsible for the transport of oxygen, leads to anemia. Mutations in the gene that codes for the hemoglobin molecule are responsible for the faulty synthesis (Mo et al. 2004). However, thalassemia does not occur as a result of a one particular mutation; rather, at least seventeen completely different mutations in different regions of the hemoglobin gene are known to cause the defective hemoglobin protein. In two people with thalassemia, different mutations can be responsible for causing the same disease.

Both Genes and Environment Make an Individual

Genes are clearly important, but they do not exist in a vacuum. Metabolic and physiological traits develop under the influence of both genes and environment, creating a unique blend expressed as a genotype-environment interaction. Genes provide only information about a phenotypic characteristic (including some diseases); the characteristics that actually develop in an individual depend on both genetic and environmental influences. As previously mentioned, genetically determined diseases such as phenylketonuria can be controlled by lifestyle—in this case, an appropriate diet.

But the most convincing evidence of the impact of environment on the development of phenotypic traits comes from the study of plants. For example, when the roots of a single plant are divided and replanted separately, some of the newly developing plants may thrive, others merely look okay, and others may barely survive. Because they originated from a single plant cut into several pieces, these plants are genetically identical. So why do they look so different? Most likely, they were planted in slightly different environments, with differences in light exposure, moisture, soil composition, or competition from other plant species. Genes only partially determined the phenotypes of the new plants: a daisy is still a daisy, but depending on the environment in which each daisy is planted, it may differ in rate of growth, final height, size and color of petals, and number of seeds.

But that is not all. *Developmental noise,* another source of variation at the phenotypic level, results from random events occurring within cells at the level of molecular rearrangements and interactions (Lewontin 2000, 36; Aranda-Anzaldo and Dent 2003). Variation in the concentration of molecules in different cells or in the distances that molecules must travel to reach their targets are among the reasons for variation in the developmental schedules of the cells of developing organs, with resulting differences in the final product. Developmental noise contributes to the expressed variation among individuals and, together with the influence of the environment, explains why two identical genotypes do not develop into two phenotypically identical organisms. Developmental noise also causes fluctuating asymmetry, producing differences between the left and right sides of bilateral structures. This is why the two sides of our faces are usually not perfectly symmetrical and the length of our fingers differs slightly between the left and right hands (Møller and Manning 2003).

Is Breast Cancer Caused by Genes?

Public health and medical students often list genes as the most important factor responsible for developing breast cancer. Although genotypes that greatly increase the risk of breast cancer have indeed been identified, only about 5 to 10 percent of all breast cancer cases can be attributed to having such genotypes (Calderon-Margalit and Paltiel 2004). This means that more than 90 percent of breast cancer cases occur in women without specific genetic causes. Mutations in genes *BRCA1* or *BRCA2* are the most common genetic factors linked with breast cancer susceptibility. An average woman in the United States has an approximately 12 percent chance that she will develop breast cancer (American Cancer Society 2012), but a woman with the *BRCA1* mutation has a 51 to 95 percent chance, and a woman with the *BRCA2* mutation a 33 to 95 percent chance that she will develop breast cancer if she lives until the age of 70 (Bermejo-Pérez, Márquez-Calderón, and Llanos-Méndez 2007). It is important to emphasize that, although the presence of breast cancer genes greatly increases the risk of breast cancer, it does not lead to a certain negative outcome. In fact, 20 to 40 percent of women having the *BRCA1* mutation will never succumb to breast cancer!

We said that both genes and environment have an influence on the development of a phenotypic trait. Is this also true in the case of breast cancer? The importance of the interaction of genes with the environment is illustrated by the way the risk of breast cancer for women who have *BRCA* mutations has increased over time (King, Marks, and Mandell 2003). For women with such mutations who were born before 1940, the risk of developing breast cancer by the age of 50 was 24 percent, but among those born after 1940 it increased to 67 percent. We are not certain what factors are responsible for this increase in risk, but particular lifestyle characteristics—especially breastfeeding a child and being physically active in the adolescent years—appear to lower the risk of the disease for women with *BRCA* mutations (King, Marks, and Mandell 2003; Jernstrom et al. 2004).

One may argue that, because appropriate lifestyle changes can reduce inherited risk, it is important to perform genetic tests to identify the mutation carriers and "prescribe" risk-reducing behaviors such as breastfeeding and physical activity. Exactly the same behaviors, however, reduce the risk of breast cancer in all women, not just in those with genetic susceptibility. So we should simplify our recommendations and make them independent of genotype!

Some Problems with Genetic Therapies

Diseases with simple, identifiable genetic causes are very rare. Cystic fibrosis is detected in 1 in 2,300 people and Huntington's disease in 1 in 10,000. Successful treatment of these diseases is very desirable, but curing them by genetic therapy or some other measure will not contribute much to the improvement of the general health status of modern populations. In 2005, just like for many years before, the National Center for Health Statistics showed that the main cause of death in the United States was heart disease. In Europe, cardiovascular disease is also the main cause of death, claiming 4.35 million lives each year (www.heartstats.org)—nearly half (49 percent) of all deaths in European countries. More specifically, 55 percent of all deaths in women and 43 percent of all deaths in men are from these diseases. Although some people may have a genetic predisposition that increases their risk of cardiovascular disease, it is well established that this is mainly a disease of lifestyle (Wu 1999). Smoking, lack of physical activity, bad dietary habits, and obesity greatly increase the risk of developing a cardiovascular disease (www.americanheart.org).

But even if there are genes that predispose some people to certain diseases, genes do not usually make it inevitable that the disease will develop. Genes for obesity will not, in most cases, make an individual obese who eats little and exercises a great deal. Genes contribute to the norm of reaction of the genotype, that is, the way an organism develops and reacts to the environment. The norm of reaction of a given genotype is impossible to predict and can only be precisely deciphered by experiments, but we can assume that two people having obesity "genes" will develop different body sizes owing to differences in their lifestyles.

The so-called *FTO* (fat mass and obesity associated) gene variants are quite common in European populations, and having each "risk allele" is associated with a 1.75 kilogram increase in body weight (Frayling et al. 2007). However, some people, despite having *FTO* alleles, have a normal body weight. In the Amish of Lancaster County, Pennsylvania—a population of European origin—people with high physical activity levels do not experience any excessive weight gain even when they have *FTO* alleles (Rampersaud et al. 2008).

Obesity increases the risk of cardiovascular disease (Timar, Sestier, and Levy 2000), but do we really believe that gene therapy is the main option for health prevention and that alleles predisposing individuals to obesity should be replaced by "normal" alleles? At present, no method for replacing faulty alleles with "healthy" alleles is in use for disease prevention, but let us consider the following hypothetical scenario. Controlling body weight through diet and physical activity clearly requires more effort from a "patient" than undergoing genetic manipulation. Lifestyle changes, however, contribute not only to keeping body weight under control but also provide benefits in many other aspects of health. Physical activity lowers the risk of not only cardiovascular disease but also other prevalent modern diseases, including adult onset diabetes, breast and colon cancers, and osteoporosis (Byers et al. 2002; Oguma and Shinoda-Tagawa 2004). In general, physically active people are in better shape, are less likely to suffer from stress and depression, and are more independent in old age. A person who has undergone genetic therapy that removes the worry of gaining weight will be less likely to exercise regularly and, as a consequence, will have an increased risk of other diseases. Are we going to fix the patient's genes for those diseases as well?

George Williams and Randolph Nesse (1991) point to yet another problem associated with gene therapy. The genes that cause an increased predisposition to a disease may be associated with benefits for some

237

other aspect of physiology or health. "If we naively assume that ridding the population of the genetic capability for Alzheimer's disease is an unconditionally desirable goal, we might incidentally eliminate unsuspected benefits" (Williams and Nesse 1991, 13). Such a situation where one gene affects many physiological or morphological traits is very common, and geneticists call it *pleiotropy*. *Antagonistic pleiotropy* occurs when the same gene has a positive effect on one trait but a detrimental effect on another.

Long Life in Perfect Health

Is perfect health possible to achieve and maintain at all? Clearly there are individuals who live for a long time without major health complaints. Are they just lucky, do they have perfect genes, or have they discovered a secret recipe for health and longevity? Newspaper interviews with such people, often conducted when they turn 100 years old, reveal that they are almost always convinced that they know the reason for their own good health and long life: "I have never eaten meat," "I have never drunk vodka," "I always have a glass of brandy in the afternoon," and so on. Jeanne Louise Calment, a French woman with the longest confirmed life to date, lived until the age of 122. She attributed her longevity to olive oil, port wine, and eating almost 1 kilogram of chocolate every week. A Polish newspaper recently described a study on centenarians in Poland that revealed that they are usually in a good mood and are generally happy, which suggested to the investigators that psychological predisposition is important for achieving longevity. Unfortunately, these results and the advice from long-lived individuals are without any application for the rest of the population. Cause and effect are often impossible to disentangle: it is equally likely that these Polish Methuselahs are happy because they have lived such long lives as it is that their longevity resulted from having a positive outlook on life.

The Rothberg Institute in the United States has taken a different approach to discovering the reasons behind good health and longevity. This nonprofit organization recently announced that it would sequence the genomes from 100 people aged 100 or older to identify the genes responsible for living long and healthy lives. Participation in this "Methuselah Project" (www.longlifegenes.org) requires filling a collection cup with saliva for genetic analysis and completing a three-page health history questionnaire. The questionnaire includes a very long list of possible diseases—but surprisingly few questions about lifestyle. An ABC News Internet article about this project was illustrated with a picture of George Burns, the comedian and actor, holding a cigar.

Burns, who died shortly after his 100th birthday, famously claimed to have smoked 10 to 15 cigars per day since he was 14 years old. Choosing him as an example of longevity promotes the misconception that living a healthy and long life is determined by our genes, not by lifestyle—not even by the combined effects of genotype and lifestyle. This perception appears to be at least partially shared by the scientists of the Rothberg Institute, judging by their lack of interest in lifestyle characteristics. The Methuselah Project does not expect to find a single gene but rather a bunch of genes, some reducing the risks of various diseases, others slowing the aging process. The researchers are convinced that a list of life- and health-enhancing genes will emerge.

But if genes are not the most important factor in health and longevity, can genuinely useful advice be given to people for significantly lowering their risk of disease? Of course, we are not talking about genetically determined diseases but about lifestyle-related and metabolic diseases that most often develop in middle and old age such as cardiovascular diseases, adult onset diabetes, certain types of cancer such as breast cancer, and osteoporosis. These are major public health problems in many countries (Timar, Sestier, and Levy 2000; Bray, McCarron, and Parkin 2004; Woolf 2006; Coleman et al. 2008).

Sometimes the advice can be really simple. Not smoking dramatically reduces the likelihood of lung cancer (Berkson 1955). If one does not smoke and is not exposed to secondhand smoke, and if one's work environment does not require contact with certain substances such as asbestos, one generally need not worry about developing that type of cancer. In England, for example, approximately 90 percent of lung cancer deaths are caused by smoking (Cancer Research UK 2012). Prevention of many other diseases is, however, much more complex. In particular, measures taken to reduce the risk of one disease should not lead to an increased risk of another.

Is Breast Cancer Preventable?

Throughout this book I have discussed breast cancer in women. Now I would like to review selected aspects of this topic in view of understanding whether we can successfully reduce the risk of this disease. Breast cancer was, most likely, very rare during human evolutionary history (Eaton and Eaton III 1999). Age at menarche was late, and women had several children and long nursing periods for each child, which amounted to a very long cumulative lifetime nursing time. All these factors are associated with a low risk of breast cancer. Weight gain and a high body weight after menopause (Ballard-Barbash et al. 1990; Huang et al. 1997), which increase the risk of breast cancer in contemporary populations, were also not a problem during evolutionary times. High levels of physical activity, which further decrease the risk, were universal features of the Paleolithic lifestyle (Cordain et al. 1998). Given this low risk, it is no wonder that the human species has not evolved adaptations to protect women against the disease.

Practically all the risk factors that contribute to an increased risk of breast cancer are related to levels of ovarian steroid hormones, especially estradiol (Ellison 1999; Jasienska, Thune, and Ellison 2000; Blamey et al. 2004) but possibly also progesterone (Wiebe 2006). That there is a connection between these hormones and breast cancer was suggested as early as 1896, when Scottish surgeon George Thomas Beatson announced that he had successfully treated premenopausal breast cancer by removal of the ovaries (Love and Philips 2002), thereby removing the major source of reproductive hormones. More recently, ovarian steroid hormones have again been singled out as an important factor in both the development and the prognosis of breast cancer (Henderson, Ross, and Bernstein 1988; Key and Pike 1988; Bernstein 2002; Eliassen et al. 2006).

At the cellular level, both estradiol and progesterone act to regulate mitosis and cellular proliferation (Soto and Sormenschein 1987; Clarke and Sutherland 1990; Henderson and Feigelson 2000). Estradiol acts as a potent mitogen (i.e., a factor stimulating cell division) in target tissues, including the cells of the breast. Progesterone also stimulates mitotic activity in both normal and neoplastic (tumorous) breast tissue. The ability of estrogens and progesterone to stimulate breast cells to divide and multiply is especially useful during pregnancy when the breasts are preparing for milk production. However, these hormones act on breast cells not only during pregnancy but also during menstrual cycles and even after menopause. A high number of mitotic divisions in healthy breast tissue increases the probability of mutations, which may lead to the development of a neoplastic tumor. Once a neoplasm has formed, its growth is faster when high concentrations of steroid hormones are available to stimulate mitotic divisions.

240

Another mechanism that explains why estrogen increases the risk of breast cancer is related to the metabolism of estrogen (Coyle 2008). Before estrogen can be removed from the body, it is metabolized to so-called hydroxylated forms, which are further chemically changed by some enzymes. Hydroxylated estrogens are dangerous molecules because they may directly damage the DNA. Hydroxylated estrogens are usually quickly neutralized, but in some situations their levels can become high and thus induce substantial DNA damage.

Exposure to high levels of ovarian steroid hormones, especially in the long term, is thought to play a role in breast cancer (Bernstein and Ross 1993). As already mentioned, most factors that are well known to increase the risk of this disease are linked to high levels of hormones. Early menarche, an established risk factor, operates via levels of ovarian hormones through at least two pathways. First, menstrual cycles early in life increases the period during which breasts are exposed to steroid hormones. Second, women with early menarche have higher levels of ovarian hormones in subsequent menstrual cycles (Apter 1996). Thus, early menarche increases both the duration and intensity of exposure to levels of circulating steroids.

Most other factors that either increase or decrease the risk of breast cancer also operate through the levels of hormones. Breastfeeding, when long, frequent, and especially when done by mothers with poor nutritional status, can completely block the production of ovarian steroids. If a woman nurses several consecutive children, the cumulative lifetime reduction in steroid levels can be quite substantial. Physical activity also reduces the levels of steroids during menstrual cycles and prevents weight gain. A high body weight after the menopause, when the ovaries do not produce hormones anymore, increases the risk of breast cancer because fat tissue has a high level of aromatase, an enzyme that converts hormones produced by the adrenal gland to estradiol.

Natural selection, as mentioned earlier, did not produce adaptations that protect women from breast cancer because this disease was very rare during our evolutionary past. Additional important factors also worked against the evolution of potential anticancer adaptations. As discussed before, the levels of estradiol and progesterone have a direct positive influence on fertility. Their high levels enhance fertility during the reproductive years; thus, even if they significantly increase mortality later on, high levels were selected *for* by natural selection. The postreproductive cost paid by individual women in this case is immaterial. That these hormones influence breast tissue is not an accident: by stimulating cell divisions, they promote the growth of breast tissue during maturation, pregnancy, and in preparation for lactation. The susceptibility of breast tissue to steroid hormones could not have been changed by natural selection because of these hormones' importance for reproduction, even though hormonal exposure may later lead to breast cancer.

Given that natural selection (for very good reasons) has done nothing to protect women against breast cancer, what can be done to reduce its incidence in contemporary populations? Theoretically, many risk factors can be targeted, but in practice things are much more complicated. For example, lowering the age of menopause would lead to a shorter exposure to steroid levels, but it would be an unlikely choice for most women. In fact, many women today decide to extend the period of steroid exposure by taking hormonal replacement therapy. Changing some characteristics of the lifestyle, such as having more children and prolonging the period of breastfeeding, would directly lead to lowering the lifetime steroid exposure and thus the risk of breast cancer but is too impractical for the majority of women to be suggested as a public health intervention. In addition, we should keep in mind that there are other health aspects that should not be overlooked when manipulations of the lifetime levels of hormones are attempted.

What Can Be Done about High Levels of Hormones?

Genes together with conditions during fetal development and childhood development influence the level of steroid hormones in adult women. High levels of hormones result from having good fetal and childhood nutritional conditions. The same factors increase the risk of breast cancer, possibly operating via increased levels of hormones. Women from the Nurses' Health Study, a large epidemiologic study that has observed over 200,000 female registered nurses in the United States since the mid-1970s, had an approximately 30 percent lower risk of breast cancer if their birth weight was between 3,000 and 3,499 grams, compared with nurses whose birth weight was 4,000 grams or more (Michels et al. 1996). Nurses who weighed less than 2,500 grams at birth had their risk of breast cancer reduced by almost 50 percent.

The impact of childhood conditions on the subsequent risk of breast cancer in women is suggested by a Norwegian study that showed a positive relationship between breast cancer and adult height (Tretli 1989), a good indicator of nutritional conditions during childhood. The importance of childhood conditions is also implied by another study from Norway where the incidence of breast cancer was lower than expected among the women who went through puberty during World War II (Tretli and Gaard 1996). The rate of breast cancer in Norway as analyzed from 1906 grew for women who went through puberty before the war, dropped by 13 percent for those who matured during the war, and increased again for the women who matured after the war. It seems that poor nutrition during fetal and childhood developments leads to a lower risk of breast cancer in later years. But, of course, no one can propose a worsening of fetal and childhood conditions as a public health measure, even though it might lead to a lower risk of breast cancer. This is not only because of ethics considerations but also because of the trade-offs involved. Although a high birth weight is associated with an increased risk of breast cancer, low birth weight, as mentioned earlier, is related to an increased risk of several metabolic diseases.

What is left? How can breast cancer be prevented? Body weight and physical activity are other factors related to the risk of breast cancer (Ballard-Barbash 1994; Thune 2000). In over 5,000 college alumnae studied many years after graduation and categorized by their participation in organized athletics during college, there was a significantly lower incidence of breast cancer among those who had been college athletes than among nonathletes (Frisch et al. 1987). Numerous other studies have confirmed that physical activity, whether recreational or resulting from occupational work, reduces the risk of breast cancer (Fraser and Shavlik 1997; Thune et al. 1997; Rockhill et al. 1999; Friedenreich and Cust 2008). As already mentioned, a high body weight, especially after the menopause, and weight gain in adulthood increase the risk of breast cancer (Ballard-Barbash et al. 1990; Huang et al. 1997; Carmichael and Bates 2004).

Statistics from the National Cancer Institute show that the lifetime chance of having breast cancer for a woman in the United States is one in eight (National Cancer Institute 2012). Most breast cancers are diagnosed in women when they are postmenopausal. For younger women, the risk of developing breast cancer is relatively low. For example, for a thirty-year-old woman, the chance of having breast cancer in the next decade of her life is only 1 in 250 (Greaves 2001, 141).

Even though breast cancer is relatively rare before menopause, prevention should begin early in life. The reason for early prevention is the fact that breast cancer takes a very long time to develop: approximately twenty years until initial DNA damage leads to subsequent malignancy (Greaves 2001, 156). During this period, the internal environment characterized by a high concentration of steroid hormones may increase the chance that a mutation in DNA will lead to cancer.

How to Reduce the Risk of Breast Cancer: Exercise or Weight Loss?

One of the keys to breast cancer prevention is to design an environment in which the processes that may lead to cancer occur in a way less conducive for its development. In women during their reproductive years, both an increase in physical activity and a reduction in body weight would lead to a reduction in the level of ovarian steroid hormones (Ellison 2003a) and, as a consequence, a reduction in a cumulative lifetime exposure to these hormones. Yet I suggest that physical activity rather than dieting is the more advisable choice for breast cancer prevention (Jasienska, Thune, and Ellison 2000), after consideration of other health aspects of ovarian steroids.

Endogenous estrogens are necessary for protection against osteoporosis, cardiovascular disease (Alden 1989; Barrett-Connor and Bush 1991; Lieberman et al. 1994; Spencer, Morris, and Rymer 1999; Prior 2007), and probably Alzheimer's disease, although the evidence for the latter is less clear (Fillit 2002; Casadesus et al. 2008). Progesterone may also interact with estrogen or have an independent effect in providing cardiovascular and bone density benefits.

Exogenous estrogens and progesterone supplied as hormone replacement therapy (HRT) do not always have the same beneficial physiological effects as endogenous hormones. Exogenous hormones can lower the risk of osteoporosis, but their role in preventing cardiovascular disease and stroke has been recently questioned (Rossouw et al. 2002; Bath and Gray 2005; Billeci et al. 2008). In fact, HRT that consists of both estrogen and progestin not only does not confer cardiac protection but also may increase the risk of cardiovascular disease among postmenopausal women, especially during the first year of treatment (Manson et al. 2003). It should be noted that the use of HRT increases the risk of breast cancer (Ross et al. 2000; Banks et al. 2003), although the relationship is not straightforward. Not all types of hormones and their combinations lead to higher risk (Fiesch-Janys et al. 2008), and some studies have reported an increase in the risk of breast cancer only after HRT has been used for a lengthy amount of time (Garwood, Kumar, and Shim 2008).

If estrogens are needed for so many aspects of female health, is it possible that a reduction of their levels to prevent breast cancer would have potential negative consequences in other areas? For example, in theory, the suppression of ovarian estrogen levels through exercise and caloric restriction may increase the risk of coronary heart disease (assuming that even a slight reduction in estrogen levels is detrimental for cardiovascular function), but the reduction in body weight and improved body composition expected from the same behaviors may counterbalance the negative effects. Lower levels of ovarian steroids due to exercise may contribute to the loss of bone mineral density, increasing the risk of osteoporosis (Winters et al. 1996; Hillard and Nelson 2003), but weight-bearing exercise may have a counteracting positive effect (Alekel et al. 1995; Petit, Prior, and Barr 1999).

In summary, lower levels of estrogens achieved via lifestyle changes may lead to a reduced risk of breast cancer (and most likely also a reduced risk of endometrial and ovarian cancers). Although low estrogen levels due to physical activity and weight loss may lead to a higher risk of some diseases, the same behaviors have very well-documented roles in reducing the risk of these diseases (despite reduced estrogen levels). Exercise and weight loss seem to reduce the risk of both hormone-dependent cancers and cardiovascular disease. The effects of the same behaviors on lowering fecundity are not a problem because they are easily reversible. Recreational running and low body weight have been associated with reduced fecundity (Green et al. 1986; Green, Weiss, and Daling 1988), yet ovarian production of hormones returns once activity is reduced or body weight increased (Warren 1990).

When it comes to strong bones, however, the situation is a little different. Weight-bearing exercise promotes bone density *despite* its reducing of estrogens levels. Weight loss, by contrast, reduces estrogens levels and has no positive effects on bones. A combination of physical activity with caloric restriction may be recommended, especially for women who tend to gain weight or are significantly overweight. But for women of normal body weight, the reason to prefer physical activity as the method of choice for reduction of steroid levels is related to other health aspects of steroid hormones. Although weight loss alone reduces the level of ovarian hormones, exercise not only has similar effects on ovarian function but, in

245

addition, provides numerous other health benefits. Of course, there are many obstacles in implementing exercise programs for the general population. Most of these problems may have their roots in the human evolutionary tendency to conserve energy rather than to be physically active.

There are many reasons why one can be skeptical about the promises that in the future most health problems will be corrected and prevented by genetic therapies. The complex genetic background of diseases, the impossible to predict results of the interaction of genes with the environment, and the pleiotropic nature of action of many genes are probably the most important issues that must be resolved before we can be optimistic that molecular techniques will reduce the incidence of the most prevalent modern diseases. For now, old-fashioned prevention involving changes in our lifestyle is still the main option for most of us.

It is important to modify public health recommendations and ground them in evolutionary thinking and a solid knowledge of the lifestyle of our evolutionary ancestors, and we need ways of communicating these recommendations to the general public. An important point is that we cannot design recommendations that would ensure that following them carefully would result in perfect health. Evolutionary theory sends us a clear message that preserving health never had much importance in the agenda of natural selection.

CONCLUSION

Is Our Physiology Obsolete?

This book has used an evolutionary framework to explore variation in selected aspects of women's physiology and health. I believe that an evolutionary life history framework is important to take into account when designing comprehensive disease prevention programs that would work better than the ones used today. An evolutionary perspective comes with a dose of skepticism. Those who do not have training in evolution are more optimistic than evolutionary biologists are about our ability to design a perfect health preservation program. During the course of our evolution, just as in any other species, natural selection promoted adaptations to increase our reproductive success. Unfortunately, adaptation and health are not necessarily the same thing. In fact, what is adaptive in evolutionary terms may well be detrimental to our health.

We cannot afford to ignore our evolutionary legacy if we are concerned with the prevention of illness in contemporary populations. Prevention of diseases is always preferable over treatment. Prevention is never perfect, of course—it can lower the risk of diseases, but it cannot completely block their occurrence. Also, some diseases are more easily preventable than others. Our knowledge about what causes different diseases is not the same for all diseases, and some causes are easier to modify than others. But what makes prevention most difficult is that we have not evolved to live in perfect health.

Prevention is difficult because there are phenotypic trade-offs between different physiological functions and genetic antagonistic pleiotropies, which make us thrive when we are young and suffer the consequences when we are old. There are costs that must be paid, mother-fetus conflicts that cannot be resolved, and our intergenerational biological history that must be coped with; and there is our own biological history, from the time when we developed in the womb to our childhood and adulthood. Finally,

247

of course, there are our genes, which reflect our evolutionary history. All these factors affect our biological condition, our health, and our longevity. Some of these factors cannot be changed, while others are relatively easy to modify. The twenty-first century began with promises from molecular biology that fixing our genome would end most human health problems, but realistically we cannot hope that genetic therapy will significantly lower the risk of prevalent modern diseases. It will greatly improve treatment of many diseases, but we are still far away from genetic manipulation as a common option for disease prevention.

So far, and probably for a long time to come, the most effective reduction in disease risk can be achieved by individuals consciously making changes in their lifestyles. This traditional means of prevention is not error proof. Although so much knowledge has been accumulated about diet, exercise, smoking, drinking, and their relationship to human health, advising people on how to protect or improve their health is often a challenge. A major part of the problem is the lack of a comprehensive theory of human health in the disciplines concerned with health prevention.

Is the Evolutionary Perspective Relevant Today?

One may criticize my approach in this book, arguing that some of the concepts and ideas discussed here, even if potentially interesting in a historical context, are no longer relevant for the health of contemporary people. The costs of reproduction, for example, were clearly important determinants of women's health in the past. Women had higher fertility then, and socioeconomic conditions constrained their ability to alleviate the costs of reproduction via appropriate diets or a reduction in workload. For modern women, the costs of reproduction are no longer an issue—or so it seems.

First, this view is clearly Europe and United States centric, because there are many women in numerous populations around the world for whom the costs of reproduction remain highly applicable. In many countries fertility is still high, and, usually in those same countries, women often have inadequate diets and face high work demands. However, I would argue that the costs of reproduction may be an important determinant in the health of women and children even in populations with completely modern lifestyles. Low fertility and easy access to sufficient resources to cover the energetic expenses of pregnancy and breastfeeding do not shield Western women from many health dilemmas. Their body's wisdom is put to the test, just as it is for women from developing countries.

All organisms face evolutionary trade-offs in their physiology between investing in current or future reproduction, and this includes modern women. Low-birth-weight babies, as discussed in this book, may partially result from such trade-offs; that is, low birth weights occur because reproduction is costly. The costs of reproduction are expected to be felt, most of all, by women who experience low availability of metabolic energy. During evolutionary times and in populations with a traditional lifestyle, inadequate energy resulted from food shortages and the necessity of expending a high amount of energy during subsistence work. Acquiring the additional energy and nutrients needed to have a healthy child with a normal birth weight should not be a problem for modern women, but their bodies may experience an inadequate energy supply when they aim for low body weight through low-calorie diets and exercise.

Our physiology has no way of distinguishing whether a state of inadequate energy supply comes from an involuntary or voluntary cause. Of course, modern women can easily reverse a state of low energy availability, and the majority of them discontinue dieting or excessive exercise when they are pregnant. However, eating well and conserving energy only during pregnancy may not be enough, as has been clearly suggested by studies on supplementation for pregnant women in the rural Gambia. The nutritional condition of the mother *before* pregnancy seems to be more important for the child's birth weight and the mother's resumption of postpartum fertility than attempts to improve maternal and fetal conditions by overeating only during pregnancy. Therefore, the physiology of women who exercise intensely or follow calorie-restriction diets for long periods before pregnancy may respond to metabolic changes that occur in their organisms due to such behaviors as signals of poor environmental conditions. In these conditions, if pregnancy occurs, an optimal life-history strategy would be to restrict the energy flow to the developing baby to ensure that the maternal organism does not become depleted and that the mother can have more children in the future. The reduced birth weight of the baby may, therefore, result from one of the most classic trade-offs: between current and future reproduction. It does not matter that the mother may not be planning to have any more children.

Problems with low birth weight are magnified by the intergenerational impact of female nutritional status. If a woman's mother and grandmother went through periods of low nutrition, her own life-history strategy takes this family history into account. We know that the birth weight of a baby is lower when nutritional problems were present in past generations of that baby's family. We do not know how other life-history traits are affected in response to this family history, but we expect that the body would employ a strategy of being rather thrifty in using energy and nutrients. Being thrifty, in this context, means having a physiology and a metabolism that promote efficient energy storage.

Just like a high-energy diet for a mother during pregnancy may have a limited affect on her baby's weight, a high intake of some nutrients by pregnant women may also have little impact. Throughout human evolution, women could not have relied on having an additional intake of nutrients when pregnant. With dietary composition and the energy content of diets changing seasonally, enhanced intake of certain nutrients (compared with their levels from before pregnancy) was simply not possible for women who had increased nutritional requirements due to pregnancy or lactation. Natural selection had to work around these dietary constraints. For example, calcium supplementation during pregnancy did not occur throughout human evolution; thus, mechanisms evolved to ensure a calcium flow to the fetus without the mother having a high-calcium diet. Even though calcium supplements in many countries are available in every drugstore and are relatively cheap, they do not seem to improve the bones of the developing baby when taken by a pregnant mother; instead, the mother's body uses calcium from her own skeleton to support fetal bone development—the way this process has always worked in the evolutionary past. Even though women today can easily afford additional calcium intake while pregnant and do not have to deplete their own bones, the pattern of calcium metabolism in pregnancy evolved a long time ago and does not seem to open to change.

Biological trade-offs are occurring in the human organism even in our modern world. Reproduction is more important than long-term survival and health in the postreproductive years. The priority of reproduction has been built through thousands of generations of natural selection, and no amount of technological progress can change that. The body considers the prospect of reproduction in the future—regardless of whether a particular woman intends to have more children—to be more important than a current pregnancy. This is how our physiology evolved, and this is how it still works. Breast cancer, one of the major health problems of women today, is a result of trade-offs between reproduction and survival. In the past, high estrogen levels during the reproductive years, even though associated with a high risk of breast cancer in the postreproductive period, were promoted by natural selection, and they are still promoted today because they increase the chance of successful conception. It does not matter that most women today want only one or two children, and they do not need to stay ready to conceive for most of their life.

The Wisdom of the Body

The concept of the "wisdom of the body" has been a popular way of thinking about human physiology and health. This idea starts with the belief that our bodies have an innate ability to make the right choices and, in general, to function in a way that is good for us. Is the human body "wise"? The wisdom of the body has evolved to be beneficial for an individual's evolutionary fitness, but that same wisdom is not necessarily beneficial for human health. The wisdom of the body—the way that human anatomy, physiology, and metabolism work—evolved to facilitate the successful transfer of copies of the organism's genes to the next generation.

As long as maintaining health does not negatively affect the process of reproduction, we can trust that our body will behave in a way that also benefits our nonreproductive functions. However, reproduction is of such predominant importance that it is prioritized even if it means that, as a result, health suffers. In women, the energy and nutrients that were used up just by the maternal organism before reproduction now need to be shared with the growing child. Such restricted allocation is not cost-free for the maternal organism. A lack of sufficient energy may impair the mother's immune system and make infections more likely. The fetus's need for calcium depletes maternal bones and may lead to eventual osteoporosis for the mother. Pregnancy and lactation elicit changes in maternal physiology and metabolism that could have a negative, long-lasting impact on maternal health. This impact can be so severe that women with high fertility die earlier than women with lower reproductive costs. Despite women having to pay such a high price, costly reproductive processes still occur.

The human body is also easily confused. It evolved in a drastically different environment than that in which we live right now, and it cannot handle this new environment very well. It is confused by the abundance of high-energy food, lack of physical activity, strange reproductive patterns, and stressful social interactions. And a mismatch between the fetal and adult environments may further lead to poorly functioning physiology in postnatal life.

Therefore, we should not and cannot count too much on the wisdom of our body when we are attempting to maintain or improve our health. Our body will not protect us from today's most threatening diseases, especially those that come with older age such as cancers, cardiovascular diseases, and dementia. In fact, the wisdom of our bodies, fragile and confused as it is, makes us even more susceptible to such health problems. Awareness of the reasons, especially the evolutionary reasons, for this fragility and confusion helps us predict when and what health problems are likely to occur. Proper, timely predictions can help us design and implement successful preventive strategies, the most important recipes for a healthy and long life.

REFERENCES

ACKNOWLEDGMENTS

INDEX

Abbo, S., D. Shtienberg, J. Lichtenzveig, S. Lev-Yadun, and A. Gopher. 2003. The chickpea, summer cropping, and a new model for pulse domestication in the ancient Near East. *Quarterly Review of Biology* 78: 435–448.

Adams, A. M. 1995. Seasonal variations in energy balance among agriculturalists in central Mali: Compromise or adaptation? *European Journal of Clinical Nutrition* 49: 809–823.

Adams, J., and R. H. Ward. 1973. Admixture studies and detection of selection. *Science* 180: 1137–1143.

Ahlgren, M., J. Wohlfahrt, L. W. Olsen, T. I. A. Sorensen, and M. Melbye. 2007. Birth weight and risk of cancer. *Cancer* 110: 412–419.

Akum, A. E., A. J. Kuoh, J. T. Minang, B. M. Achimbom, M. J. Ahmadou, and M. Troye-Blomberg. 2005. The effect of maternal, umbilical cord and placental malaria parasitaemia on the birthweight of newborns from South-western Cameroon. *Acta Paediatrica* 94: 917–923.

Alden, J. C. 1989. Osteoporosis—a review. *Clinical Therapeutics* 11: 3–14.

Alekel, L., J. L. Clasey, P. C. Fehling, R. M. Weigel, R. A. Boileau, J. W. Erdman, et al. 1995. Contributions of exercise, body composition, and age to bone mineral density in premenopausal women. *Medicine and Science in Sports and Exercise* 27: 1477–1485.

Allali, F., S. El Aichaoui, B. Saoud, H. Maaroufi, R. Abouqal, and N. Hajjaj-Hassouni. 2006. The impact of clothing style on bone mineral density among post menopausal women in Morocco: A case-control study. *BMC Public Health* 6: 135–139.

Allen, N. E., M. S. Forrest, and T. J. Key. 2001. The association between polymorphisms in the CYP17 and 5 alpha-reductase (SRD5A2) genes and serum androgen concentrations in men. *Cancer Epidemiology Biomarkers and Prevention* 10: 185–189.

Alonso-Alvarez, C., S. Bertrand, G. Devevey, M. Gaillard, J. Prost, B. Faivre, et al. 2004. An experimental test of the dose-dependent effect of carotenoids and immune activation on sexual signals and antioxidant activity. *American Naturalist* 164: 651–659.

Alter, G., M. Manfredini, P. Nystedt, C. Campbell, J. Z. Lee, E. Ochiai, et al. 2004. Gender differences in mortality. In T. Bengtsson, C. Cameron, and J. Z. Lee, eds., *Life under Pressure: Mortality and Living Standards in Europe and Asia, 1700–1900*, 327–357. Cambridge, Mass.: MIT Press.

Alvard, M. S. 2001. Mutualistic hunting. In C. B. Stanford and H. T. Bunn, eds., *Meat-Eating and Human Evolution*, 261–278. New York: Oxford University Press.

Ambrosone, C. B., K. B. Moysich, H. Furberg, J. L. Freudenheim, E. D. Bowman, S. Ahmed, et al. 2003. CYP17 genetic polymorphism, breast cancer, and breast cancer risk factors. *Breast Cancer Research* 5: R45–R51.

American Cancer Society. 2012. Breast Cancer Facts & Figures 2011–2012. http://www.cancer.org/Research/CancerFactsFigures/BreastCancerFactsFigures/breast-cancer-facts-and-figures-2011-2012.

American College of Sports Medicine. 2005. *ACSM's Guidelines for Exercise Testing and Prescription*, 7th ed., 133–173. Baltimore: Lippincott, Williams and Wilkins.

American Heart Association. 2012. Good vs. bad cholesterol. Updated March 12. http://www.heart.org/HEARTORG/Conditions/Cholesterol/AboutCholesterol/Good-vs-Bad-Cholesterol_UCM_305561_Article.jsp.

Angel, J. L. 1984. Health as a crucial factor in the changes from hunting to developed farming in the Eastern Mediterranean. In M. N. Cohen and G. J. Armelagos, eds., *Paleopathology at the Origins of Agriculture*, 51–73. Orlando, Fla.: Academic Press.

Anonymous. 1904. The feeding of school children. *Lancet* 164: 860–862.

Anton, S. C. 2003. Natural history of *Homo erectus*. *Yearbook of Physical Anthropology* 46: 126–169.

Apicella, L. L., and A. E. Sobota. 1990. Increased risk of urinary-tract infection associated with the use of calcium supplements. *Urological Research* 18: 213–217.

Apter, D. 1996. Hormonal events during female puberty in relation to breast cancer risk. *European Journal of Cancer Prevention* 5: 476–482.

Apter, D., and R. Vihko. 1983. Early menarche, a risk factor for breast cancer, indicates early onset of ovulatory cycles. *Journal of Clinical Endocrinology and Metabolism* 57: 82–86.

Aranda-Anzaldo, A., and M. A. R. Dent. 2003. Developmental noise, ageing and cancer. *Mechanisms of Ageing and Development* 124: 711–720.

Archer, J. S. M. 1999. Relationship between estrogen, serotonin, and depression. *Menopause* 6: 71–78.

Bagga, D., S. Capone, H. J. Wang, D. Heber, M. Lill, L. Chap, et al. 1997. Dietary modulation of omega-3/omega-6 polyunsaturated fatty acid ratios in patients with breast cancer. *Journal of the National Cancer Institute* 89: 1123–1131.

Bailey, R. C., M. R. Jenike, P. T. Ellison, G. R. Bentley, A. M. Harrigan, and N. R. Peacock. 1992. The ecology of birth seasonality among agriculturalists in central Africa. *Journal of Biosocial Science* 24: 393–412.

Bakketeig, L. S., H. J. Hoffman, and E. E. Harley. 1979. Tendency to repeat gestational-age and birth-weight in successive births. *American Journal of Obstetrics and Gynecology* 135: 1086–1103.

Ballard-Barbash, R. 1994. Anthropometry and breast cancer: Body size—a moving target. *Cancer (Supplement)* 74: 1090–1100.

Ballard-Barbash, R., A. Schatzkin, P. R. Taylor, and L. L. Kahle. 1990. Association of change in body mass with breast cancer. *Cancer Research* 50: 2152–2155.

Bamshad, M., and A. G. Motulsky. 2008. Health consequences of ecogenetic variation. In S. C. Stearns and J. C. Koella, eds., *Evolution in Health and Disease*, 43–50. New York: Oxford University Press.

Banks, E., V. Beral, D. Bull, G. Reeves, J. Austoker, R. English, et al. 2003. Breast cancer and hormone-replacement therapy in the Million Women Study. *Lancet* 362: 419–427.

Barker, D. J. P. 1994. *Mothers, Babies, and Disease in Later Life.* London: BMJ Publishing.

———. 1995. Fetal origins of coronary heart disease. *British Medical Journal* 311: 171–174.

———. 1999. Intrauterine nutrition may be important—Commentary. *British Medical Journal* 318: 1477–1478.

Barlow, K. 1985. The social context of infant feeding in the Marik lakes of Papua New Guinea. In L. B. Marshall, ed., *Infant Care and Feeding in the South Pacific*, 137–154. New York: Gordon and Breach Science Publishers.

Barnett, J. B., M. N. Woods, B. Rosner, C. McCormack, L. Floyd, C. Longcope, and S. L. Gorbach. 2002. Waist-to-hip ratio, body mass index and sex hormone levels associated with breast cancer risk in premenopausal Caucasian women. *Journal of Medical Sciences* 2: 170–176.

Barrett-Connor, E., and T. L. Bush. 1991. Estrogen and coronary heart disease in women. *Journal of the American Medical Association* 265: 1861–1867.

Bateson, P. 2001. Fetal experience and good adult design. *International Journal of Epidemiology* 30: 928–934.

Bateson, P., D. Barker, T. Clutton-Brock, D. Deb, B. D'Udine, R. A. Foley, et al. 2004. Developmental plasticity and human health. *Nature* 430: 419–421.

Bath, P. M. W., and L. J. Gray. 2005. Association between hormone replacement therapy and subsequent stroke: A meta-analysis. *British Medical Journal* 330: 342–344A.

Bean, R. N. 1975. The imports of fish to Barbados in 1698. *Journal of the Barbados Museum and Historical Society* 35: 17–21.

Becker, A. E., S. K. Grinspoon, A. Klibanski, and D. B. Herzog. 1999. Current concepts—Eating disorders. *New England Journal of Medicine* 340: 1092–1098.

Beeton, M., G. U. Yule, and K. Pearson. 1900. Data for the problem of evolution in man. V. On the correlation between duration of life and the number of offspring. *Proceedings of the Royal Society of London* 67: 159–179.

Belachew, T., C. Hadley, D. Lindstrom, Y. Getachew, L. Duchateau, and P. Kolsteren. 2011. Food insecurity and age at menarche among adolescent girls in Jimma Zone Southwest Ethiopia: A longitudinal study. *Reproductive Biology and Endocrinology* 9: 125–132.

Bell, R. J., S. M. Palma, and J. M. Lumley. 1995. The effect of vigorous exercise during pregnancy on birth-weight. *Australian and New Zealand Journal of Obstetrics and Gynaecology* 35: 46–51.

Benefice, E., K. Simondon, and R. M. Malina. 1996. Physical activity patterns and anthropometric changes in Senegalese women observed over a complete seasonal cycle. *American Journal of Human Biology* 8: 251–261.

Bentley, G. R. 1985. Hunter-gatherer energetics and fertility: A reassessment of the !Kung San. *Human Ecology* 13: 79–109.

———. 1994. Do hormonal contraceptives ignore human biological variation and evolution? *Annals of New York Academy of Science* 709: 201–203.

Bentley, G. R., T. Goldberg, and G. Jasienska. 1993. The fertility of agricultural and non-agricultural traditional societies. *Population Studies* 47: 269–281.

Bentley, G. R., A. M. Harrigan, and P. T. Ellison. 1998. Dietary composition and ovarian function among Lese horticulturalist women of the Ituri Forest, Democratic Republic of Congo. *European Journal of Clinical Nutrition* 52: 261–270.

Bentley, G. R., G. Jasienska, and T. Goldberg. 1993. Is the fertility of agriculturalists higher than that of nonagriculturalists? *Current Anthropology* 34: 778–785.

Benzie, I. F. F. 2003. Evolution of dietary antioxidants. *Comparative Biochemistry and Physiology A—Molecular and Integrative Physiology* 136: 113–126.

Beral, V., D. Bull, R. Doll, R. Peto, G. Reeves, C. La Vecchia, et al. 2002. Breast cancer and breastfeeding: Collaborative reanalysis of individual data from 47 epidemiological studies in 30 countries, including 50302 women with breast cancer and 96973 women without the disease. *Lancet* 360: 187–195.

Bererhi, H., N. Kolhoff, A. Constable, and S. P. Nielsen. 1996. Multiparity and bone mass. *British Journal of Obstetrics and Gynaecology* 103: 818–821.

Bergman-Jungestrom, M., M. Gentile, A. C. Lundin, and S. Wingren. 1999. Association between CYP17 gene polymorphism and risk of breast cancer in young women. *International Journal of Cancer* 84: 350–353.

Bergmann, B. R. 1986. *Saving Our Children from Poverty: What the United States Can Learn from France.* New York: Russell Sage Foundation.

Berkson, J. 1955. The statistical study of association between smoking and lung cancer. *Proceedings of the Staff Meetings of the Mayo Clinic* 30: 319–348.

Bermejo-Pérez, M. J., S. Márquez-Calderón, and A. Llanos-Méndez. 2007. Effectiveness of preventive interventions in BRCA1/2 gene mutation carriers: A systematic review. *International Journal of Cancer* 121: 225–231.

Bernstein, L. 2002. Epidemiology of endocrine-related risk factors for breast cancer. *Journal of Mammary Gland Biology and Neoplasia* 7: 3–15.

Bernstein, L., and R. K. Ross. 1993. Endogenous hormones and breast cancer risk. *Epidemiological Reviews* 15: 48–65.

Bernstein, L., J. M. Yuan, R. K. Ross, M. C. Pike, R. Hanisch, R. Lobo, et al. 1990. Serum hormone levels in pre-menopausal Chinese women in Shanghai and white women in Los Angeles—Results from two breast cancer case-control studies. *Cancer Causes and Control* 1: 51–58.

Bertrand, S., C. Alonso-Alvarez, G. Devevey, B. Faivre, J. Prost, and G. Sorci. 2006. Carotenoids modulate the trade-off between egg production and resistance to oxidative stress in zebra finches. *Oecologia* 147: 576–584.

Bethea, C. L., N. Z. Lu, C. Gundlah, and J. M. Streicher. 2002. Diverse actions of ovarian steroids in the serotonin neural system. *Frontiers in Neuroendocrinology* 23: 41–100.

Billeci, A. M. R., M. Paciaroni, V. Caso, and G. Agnelli. 2008. Hormone replacement therapy and stroke. *Current Vascular Pharmacology* 6: 112–123.

Bin Li, Z., S. Yin Ho, W. Man Chan, K. Sang Ho, M. Pik Li, G. M. Leung, and T. Hing Lam. 2004. Obesity and depressive symptoms in Chinese elderly. *International Journal of Geriatric Psychiatry* 19: 68–74.

Bisdee, J., W. James, and M. Shaw. 1989. Changes in energy expenditure during the menstrual cycle. *British Journal of Nutrition* 61: 187–199.

Black, A. J., J. Topping, B. Durham, R. G. Farquharson, and W. D. Fraser. 2000. A detailed assessment of alterations in bone turnover, calcium homeostasis, and bone density in normal pregnancy. *Journal of Bone and Mineral Research* 15: 557–563.

Blackburn, S, and D Loper. 1992. *Maternal, Fetal, and Neonatal Physiology: A Clinical Perspective.* Philadelphia: W. B. Saunders.

Blamey, R., J. Collins, P. G. Crosignani, E. Diczfalusy, L. A. J. Heinemann, C. La Vecchia, et al. 2004. Hormones and breast cancer. *Human Reproduction Update* 10: 281–293.

Bledsoe, R. E., M. T. O'Rourke, and P. T. Ellison. 1990. Characterization of progesterone profiles of recreational runners [abstract]. *American Journal of Physical Anthropology* 81: 195–196.

Blonigen, B. 2004. A re-examination of the slave diet. Senior thesis, Departments of History and Nutrition, College of St. Benedict/St. John's University.

Bogin, B. 1999. *Patterns of Human Growth.* Cambridge: Cambridge University Press.

———. 2001. *The Growth of Humanity.* New York: Wiley-Liss.

Borgerhoff Mulder, M. 1988. Kipsigis Bridewealth Payments. In L. Betzig, M. Borgerhoff Mulder, and P. Turke, eds., *Human Reproductive Behavior,* 65–82. Cambridge: Cambridge University Press.

Both, M. I., M. A. Overvest, M. F. Wildhagen, J. Golding, and H. I. J. Wildschut. 2010. The association of daily physical activity and birth outcome: A population-based cohort study. *European Journal of Epidemiology* 25: 421–429.

Bramble, D. M., and D. E. Lieberman. 2004. Endurance running and the evolution of *Homo. Nature* 432: 345–352.

Bray, F., P. McCarron, and D. M. Parkin. 2004. The changing global patterns of female breast cancer incidence and mortality. *Breast Cancer Research* 6: 229–239.

Bribiescas, R. G. 2001. Reproductive physiology of the human male: An evolutionary and life history perspective. In P. T. Ellison, ed., *Reproductive Ecology and Human Evolution,* 107–136. New York: Aldine de Gruyter.

———. 2006. *Men: Evolutionary and Life History.* Cambridge, Mass.: Harvard University Press.

Bribiescas, R. G., and P. T. Ellison. 2008. How hormones mediate trade-offs in human health and disease. In S. C. Stearns and J. C. Koella, eds., *Evolution in Health and Disease,* 77–93. New York: Oxford University Press.

Bronstein, M. N., R. P. Mak, and J. C. King. 1996. Unexpected relationship between fat mass and basal metabolic rate in pregnant women. *British Journal of Nutrition* 75: 659–668.

Broocks, A., K. M. Pirke, U. Schweiger, R. J. Tuschl, R. G. Laessle, T. Strowitzki, et al. 1990. Cyclic ovarian function in recreational athletes. *Journal of Applied Physiology* 68: 2083–2086.

Brooks, A. A., M. R. Johnson, P. J. Steer, M. E. Pawson, and H. I. Abdalla. 1995. Birth-weight—Nature or nurture. *Early Human Development* 42: 29–35.

Brown, J. E., S. A. Kaye, and A. R. Folsom. 1992. Parity-related weight change in women. *International Journal of Obesity* 16: 627–631.

Brown, P. J. 1991. Culture and the evolution of obesity. *Human Nature* 2: 31–51.

Bruning, P. F., J. M. G. Bonfrer, A. A. M. Hart, P. A. H. Van Noord, H. Van Der Hoeven, H. J. A. Collette, et al. 1992. Body measurements, estrogen availability and the risk of human breast cancer: A case-control study. *International Journal of Cancer* 51: 14–19.

Bucksch, J., and W. Schlicht. 2006. Health-enhancing physical activity and the prevention of chronic diseases—An epidemiological review. *Sozial und Praventivmedizin* 51: 281–301.

Bullen, B. A., G. S. Skrinar, I. Z. Beitins, G. von Mering, B. A. Turnbull, and J. W. McArthur. 1985. Induction of menstrual disorders by strenuous exercise in untrained women. *New England Journal of Medicine* 312: 1349–1353.

Bullough, R. C., C. A. Gillette, M. A. Harris, and C. L. Melby. 1995. Interaction of acute changes in exercise energy expenditure and energy intake on resting metabolic rate. *American Journal of Clinical Nutrition* 61: 473–481.

Burke, C. M., R. C. Bullough, and C. L. Melby. 1993. Resting metabolic rate and postprandial thermogenesis by level of aerobic fitness in young women. *European Journal of Clinical Nutrition* 47: 575–585.

Burroughs, W. J. 2005. *Climate Change in Prehistory: The End of the Reign of Chaos.* Cambridge: Cambridge University Press.

Butte, N. F., J. M. Hopkinson, N. Mehta, J. K. Moon, and E. O. Smith. 1999. Adjustments in energy expenditure and substrate utilization during late pregnancy and lactation. *American Journal of Clinical Nutrition* 69: 299–307.

Butte, N. F., and J. C. King. 2005. Energy requirements during pregnancy and lactation. *Public Health Nutrition* 8: 1010–1027.

Butte, N. F., W. W. Wong, and J. M. Hopkinson. 2001. Energy requirements of lactating women derived from doubly labeled water and milk energy output. *Journal of Nutrition* 131: 53–58.

Byers, T., M. Nestle, A. McTiernan, C. Doyle, A. Currie-Williams, T. Gansler, and M. Thun. 2002. American Cancer Society guidelines on nutrition and physical activity for cancer prevention: Reducing the risk of cancer with healthy food choices and physical activity. *CA: A Cancer Journal for Clinicians* 52: 92–119.

Cabral, H., L. E. Fried, S. Levenson, H. Amaro, and B. Zuckerman. 1990. Foreign-born and US-born black women: Differences in health behaviors and birth outcomes. *American Journal of Public Health* 80: 70–72.

Cade, J. E., V. J. Burley, and D. C. Greenwood. 2007. Dietary fibre and risk of breast cancer in the UK Women's Cohort Study. *International Journal of Epidemiology* 36: 431–438.

Calderon-Margalit, R., and O. Paltiel. 2004. Prevention of breast cancer in women who carry BRCA1 or BRCA2 mutations: A critical review of the literature. *International Journal of Cancer* 112: 357–364.

Campbell, C., and J. Z. Lee. 2004. Mortality and household in seven Liaodong populations, 1749–1909. In T. Bengtsson, C. Cameron, and J. Lee, eds., *Life under Pressure: Mortality and Living Standards in Europe and Asia, 1700–1900*, 293–324. Cambridge, Mass.: MIT Press.

Campbell, J. 1984. Work, pregnancy, and infant-mortality among Southern slaves. *Journal of Interdisciplinary History* 14: 793–812.

Campbell, K. L., and J. W. Wood. 1988. Fertility in traditional societies. In P. Diggory, S. Teper, and M. Potts, eds., *Natural Human Fertility*, 39–69. London: Macmillan.

Cancer Research UK. 2012. Lung cancer and smoking. Cancer Research UK, News & Resources, April 19. http://info.cancerresearchuk.org/cancerstats/types /lung.

Cannon, W. B. 1932. *The Wisdom of the Body*. New York: W. W. Norton.

Carlson, E. D. 1984. Social determinants of low birth weight in a high risk population. *Demography* 21: 207–216.

Carmichael, A. R., and T. Bates. 2004. Obesity and breast cancer: A review of the literature. *Breast* 13: 85–92.

Casadesus, G., R. K. Rolston, K. M. Webber, C. S. Atwood, R. L. Bowen, G. Perry, and M. A. Smith. 2008. Menopause, estrogen, and gonadotropins in Alzheimer's disease. *Advances in Clinical Chemistry* 45: 139–153.

Casanueva, E., and F. E. Viteri. 2003. Iron and oxidative stress in pregnancy. *Journal of Nutrition* 133 (Suppl. 2): 1700S–1708S.

Cashdan, E. 1998. Adaptiveness of food learning and food aversions in children. *Social Science Information* 37: 613–632.

Catalano, P. M., E. D. Tyzbir, N. M. Roman, S. B. Amini, and E. A. H. Sims. 1991. Longitudinal changes in insulin release and insulin resistance in nonobese pregnant women. *American Journal of Obstetrics and Gynecology* 165: 1667–1672.

Ceesay, S. M. 1997. Effects on birth weight and perinatal mortality of maternal dietary supplements in rural Gambia: 5 year randomised controlled trial. *British Medical Journal* 315: 1141.

Chaffkin, L. M., A. A. Luciano, and J. J. Peluso. 1993. The role of progesterone in regulating human granulosa cell proliferation and differentiation in vitro. *Journal of Clinical Endocrinology and Metabolism* 76: 696–700.

Chang, B. L., S. Q. L. Zheng, S. D. Isaacs, K. E. Wiley, J. D. Carpten, G. A. Hawkins, et al. 2001. Linkage and association of CYP17 gene in hereditary and sporadic prostate cancer. *International Journal of Cancer* 95: 354–359.

Chasan-Taber, L., K. R. Evenson, B. Sternfeld, and S. Kengeri. 2007. Assessment of recreational physical activity during pregnancy in epidemiologic studies of birthweight and length of gestation: Methodologic aspects. *Women and Health* 45: 85–107.

Chen, W. C., M. H. Tsai, W. Lei, W. C. Chen, C. H. Tsai, and F. J. Tsai. 2005. CYP17 and tumor necrosis factor-alpha gene polymorphisms are associated with risk of oral cancer in Chinese patients in Taiwan. *Acta Oto-Laryngologica* 125: 96–99.

Chiche, J. D., A. Cariou, and J. P. Mira. 2002. Bench-to-bedside review: Fulfilling promises of the Human Genome Project. *Critical Care* 6: 212–215.

Chippindale, A. K., J. R. Gibson, and W. R. Rice. 2001. Negative genetic correlation for adult fitness between sexes reveals ontogenetic conflict in *Drosophila*. *Proceedings of the National Academy of Sciences of the United States of America* 98: 1671–1675.

Church, T. S., Y. J. Cheng, C. P. Earnest, C. E. Barlow, L. W. Gibbons, E. L. Priest, and S. N. Blair. 2004. Exercise capacity and body composition as predictors of mortality among men with diabetes. *Diabetes Care* 27: 83–88.

Cicognani, A., R. Alessandroni, A. Pasini, P. Pirazzoli, A. Cassio, E. Barbieri, and E. Cacciari. 2002. Low birth weight for gestational age and subsequent male gonadal function. *Journal of Pediatrics* 141: 376–380.

Clapp, J. F. 2000. Exercise during pregnancy—A clinical update. *Clinics in Sports Medicine* 19: 273–286.

Clark, A. M., W. Ledger, C. Galletly, L. Tomlinson, F. Blaney, X. Wang, and R. J. Norman. 1995. Weight-loss results in significant improvement in pregnancy and ovulation rates in anovulatory obese women. *Human Reproduction* 10: 2705–2712.

Clarke, C. L., and R. L. Sutherland. 1990. Progestin regulation of cellular proliferation. *Endocrine Reviews* 11: 266–301.

Clayton, F., J. Sealy, and S. Pfeiffer. 2006. Weaning age among foragers at Matjes River Rock Shelter, South Africa, from stable nitrogen and carbon isotope analyses. *American Journal of Physical Anthropology* 129: 311–317.

Cliggett, L. 2005. *Grains from Grass: Aging, Gender, and Famine in Rural Africa.* Ithaca, N.Y.: Cornell University Press.

Cody, C. A. 1977. A note on changing patterns of slave fertility in the South Carolina Rice District, 1735–1865. *Southern Studies* 16: 457–463.

Cohen, M. N., and G. J. Armelagos. 1984. Paleopathology at the origins of agriculture: Editors' summation. In M. N. Cohen and G. J. Armelagos, eds., *Paleopathology at the Origins of Agriculture*, 585–601. Orlando, Fla.: Academic Press.

Cole, A. H., P. A. Ibeziako, and E. A. Bamgboye. 1989. Basal metabolic rate and energy expenditure of pregnant Nigerian women. *British Journal of Nutrition* 62: 631–638.

Coleman, M. P., M. Quaresma, F. Berrino, J. M. Lutz, R. De Angelis, R. Capocaccia, et al. 2008. Cancer survival in five continents: A worldwide population-based study (CONCORD). *Lancet Oncology* 9: 730–756.

Colen, C. G., A. T. Geronimus, J. Bound, and S. A. James. 2006. Maternal upward socioeconomic mobility and black-white disparities in infant birthweight. *American Journal of Public Health* 96: 2032–2039.

Collins, J. W., R. J. David, A. Handler, S. Wall, and S. Andes. 2004. Very low birth-weight in African American infants: The role of maternal exposure to interpersonal racial discrimination. *American Journal of Public Health* 94: 2132–2138.

Collins, J. W., S. Y. Wu, and R. J. David. 2002. Differing intergenerational birth weights among the descendants of US-born and foreign-born whites and African Americans in Illinois. *American Journal of Epidemiology* 155: 210–216.

Conton, L. 1985. Social, economic, and ecological parameters of infant feeding in Usino, Papua New Guinea. In L. Marshall, ed., *Infant Care and Feeding in the South Pacific*, 97–120. New York: Gordon and Breach Science Publishers.

Cordain, L., S. B. Eaton, A. Sebastian, N. Mann, S. Lindeberg, B. A. Watkins, et al. 2005. Origins and evolution of the Western diet: Health implications for the 21st century. *American Journal of Clinical Nutrition* 81: 341–354.

Cordain, L., R. W. Gotshall, and S. B. Eaton. 1997. Evolutionary aspects of exercise. In A. Simopoulos, ed., *Nutrition and Fitness: Evolutionary Aspects, Children's Health, Programs and Policies*, 49–60. Basel, Switzerland: Karger.

Cordain, L., R. W. Gotshall, S. B. Eaton, and S. B. Eaton III. 1998. Physical activity, energy expenditure and fitness: An evolutionary perspective. *International Journal of Sports Medicine* 19: 328–335.

Cordain, L., J. B. Miller, S. B. Eaton, N. Mann, S. H. A. Holt, and J. D. Speth. 2000. Plant-animal subsistence ratios and macronutrient energy estimations in worldwide hunter-gatherer diets. *American Journal of Clinical Nutrition* 71: 682–692.

Cosminski, S. 1985. Infant feeding practices in rural Kenya. In V. Hull and M. Simpson, eds., *Breastfeeding, Child Health and Child Spacing: Cross-Cultural Perspectives*, 35–54. London: Croom Helm.

Costa, D. L. 2004. Race and pregnancy outcomes in the twentieth century: A long-term comparison. *Journal of Economic History* 64: 1056–1086.

Coyle, Y. 2008. Physical activity as a negative modulator of estrogen-induced breast cancer. *Cancer Causes and Control* 19: 1021–1029.

Creighton, D. L., A. L. Morgan, D. Boardley, and P. G. Brolinson. 2001. Weight-bearing exercise and markers of bone turnover in female athletes. *Journal of Applied Physiology* 90: 565–570.

Criqui, M. H., and B. L. Ringel. 1994. Does diet or alcohol explain the French paradox? *Lancet* 344: 1719–1723.

Crisp, A. H., and B. McGuiness. 1976. Jolly fat—Relation between obesity and psychoneurosis in general population. *British Medical Journal* 1: 7–9.

Cui, J. S., A. B. Spurdle, M. C. Southey, G. S. Dite, D. J. Venter, M. R. E. McCredie, et al. 2003. Regressive logistic and proportional hazards disease models for within-family analyses of measured genotypes, with application to a CYP17 polymorphism and breast cancer. *Genetic Epidemiology* 24: 161–172.

Cummings, S. R., M. C. Nevitt, W. S. Browner, K. Stone, K. M. Fox, K. E. Ensrud, et al. 1995. Risk-factors for hip fracture in white women. *New England Journal of Medicine* 332: 767–773.

Curtis, V., C. J. K. Henry, E. Birch, and A. Ghusain Choueiri. 1996. Intraindividual variation in the basal metabolic rate of women: Effect of the menstrual cycle. *American Journal of Human Biology* 8: 631–639.

Dalkiranis, A., T. Patsanas, S. K. Papadopoulou, I. Gissis, O. Denda, and A. Mylo-nas. 2006. Bone mineral density in fin swimmers. *Journal of Human Movement Studies* 50: 19–28.

David, R. J., and J. W. Collins. 1997. Differing birth weight among infants of U.S.-born blacks, African-born blacks, and U.S.-born whites. *New England Journal of Medicine* 337: 1209–1214.

Davies, M. J., and R. J. Norman. 2002. Programming and reproductive functioning. *Trends in Endocrinology and Metabolism* 13: 386–392.

De Aloysio, D., P. Di Donato, N. A. Giulini, B. Modena, G. Cicchetti, G. Comitini, et al. 2002. Risk of low bone density in women attending menopause clinics in Italy. *Maturitas* 42: 105–111.

de Bruin, J. P., M. Dorland, H. W. Bruinse, W. Spliet, P. G. J. Nikkels, and E. R. Te Velde. 1998. Fetal growth retardation as a cause of impaired ovarian development. *Early Human Development* 51: 39–46.

DeBusk, R. F., U. Stenestrand, M. Sheehan, and W. L. Haskell. 1990. Training effects of long versus short bouts of exercise in healthy subjects. *American Journal of Cardiology* 65: 1010–1013.

de Heinzelin, J., J. D. Clark, T. White, W. Hart, P. Renne, G. WoldeGabriel, et al. 1999. Environment and behavior of 2.5-million-year-old Bouri hominids. *Science* 284: 625–629.

De Pergola, G., S. Maldera, M. Tartagni, N. Pannacciulli, G. Loverro, and R. Giorgino. 2006. Inhibitory effect of obesity on gonadotropin, estradiol, and inhibin B levels in fertile women. *Obesity* 14: 1954–1960.

De Souza, M. J., B. E. Miller, A. B. Loucks, A. A. Luciano, L. S. Pescatello, C. G. Campbell, and B. L. Lasley. 1998. High frequency of luteal phase deficiency and anovulation in recreational women runners: Blunted elevation in follicle-stimulating hormone observed during luteal-follicular transition. *Journal of Clinical Endocrinology and Metabolism* 83: 4220–4232.

Diamanti-Kandarakis, E., M. I. Bartzis, E. D. Zapanti, G. G. Spina, F. A. Filandra, T. C. Tsianateli, et al. 1999. Polymorphism T → C (−34 bp) of gene CYP17 promoter in Greek patients with polycystic ovary syndrome. *Fertility and Sterility* 71: 431–435.

Diamond, J. 1997. *Guns, Germs, and Steel: The Fates of Human Societies.* New York: W. W. Norton.

Diamond, J., and P. Bellwood. 2003. Farmers and their languages: The first expansions. *Science* 300: 597–603.

Dickemann, M. 1979. The ecology of mating systems in hypergynous dowry societies. *Social Science Information* 18: 163–195.

———. 1981. Paternal confidence and dowry competition: A biocultural analysis of purdah. In R. D. Alexander and D. W. Tinkle, eds., *Natural Selection and Social Behavior,* 417–438. New York: Chiron Press.

Dickey, R. P., T. T. Olar, S. N. Taylor, D. N. Curole, and E. M. Matulich. 1993. Relationship of endometrial thickness and pattern to fecundity in ovulation induction cycles: Effect of clomiphene citrate alone and with human menopausal gonadotropin. *Fertility and Sterility* 59: 756–760.

Dirks, R. 1978. Resource fluctuations and competitive transformation in West Indian slave societies. In D. Laughlin and I. A. Brody, eds., *Extinction and Survival in Human Populations*, 122–180. New York: Columbia University Press.

Dixon, J. B., M. F. Dixon, and P. E. O'Brien. 2003. Depression in association with severe obesity—Changes with weight loss. *Archives of Internal Medicine* 163: 2058–2065.

Dixon, R. A. 2004. Phytoestrogens. *Annual Review of Plant Biology* 55: 225–261.

Doblhammer, G., and J. Oeppen. 2003. Reproduction and longevity among the British peerage: The effect of frailty and health selection. *Proceedings of the Royal Society of London B, Biological Sciences* 270: 1541–1547.

Dole, N., D. A. Savitz, A. M. Siega-Riz, I. Hertz-Picciotto, M. J. McMahon, and P. Buelkens. 2004. Psychosocial factors and preterm birth among African American and white women in central North Carolina. *American Journal of Public Health* 94: 1358–1365.

Dolezal, B. A. , and J. A. Potteiger. 1998. Concurrent resistance and endurance training influence basal metabolic rate in nondieting individuals. *Journal of Applied Physiology* 85: 695–700.

Dong, C., L. E. Sanchez, and R. A. Price. 2004. Relationship of obesity to depression: A family-based study. *International Journal of Obesity* 28: 790–795.

Dorkenoo, E. 1994. *Cutting the Rose: Female Genital Mutilation, the Practice and Its Prevention*. London: Minority Rights Group.

Dorozynski, A. 2003. France offers E800 reward for each new baby. *British Medical Journal* 326: 1002.

Draper, E. S., K. R. Abrams, and M. Clarke. 1995. Fall in birth weight of 3rd-generation Asian infants. *British Medical Journal* 311: 876–876.

Dribe, M. 2004. Long-term effects of childbearing on mortality: Evidence from pre-industrial Sweden. *Population Studies—A Journal of Demography* 58: 297–310.

Drinkwater, B. L., and C. H. Chestnut. 1991. Bone-density changes during pregnancy and lactation in active women—A longitudinal study. *Bone and Mineral* 14: 153–160.

Dufour, D. L., and M. L. Sauther. 2002. Comparative and evolutionary dimensions of the energetics of human pregnancy and lactation. *American Journal of Human Biology* 14: 584–602.

Dunn, F. L. 1968. Epidemiological factors: Health and disease in hunter-gatherers. In R. B. Lee and I. DeVore, eds., *Man the Hunter*, 221–228. Chicago: Aldine.

Dunning, A. M., C. S. Healey, P. D. P. Pharoah, N. A. Foster, J. M. Lipscombe, K. L. Redman, et al. 1998. No association between a polymorphism in the steroid metabolism gene CYP17 and risk of breast cancer. *British Journal of Cancer* 77: 2045–2047.

Durnin, J. V. G. A. 1991. Energy requirements of pregnancy. *Acta Paediatrica Scandinavica Supplement* 373: 33–42.

———. 1993. Energy requirements in human pregnancy, in human nutrition and parasitic infection. *Parasitology* 107: S169–S175.

Dwork, D. 1987. *War Is Good for Babies and Other Young Children: A History of the Infant and Child Welfare Movement in England 1898–1918.* London: Tavistock.

Eaton, S. B., and L. Cordain. 1997. Evolutionary aspects of diet: Old genes, new fuels—Nutritional changes since agriculture. *World Review of Nutrition and Dietetics* 81: 26–37.

Eaton, S. B., and S. B. Eaton III. 1999. Breast cancer in evolutionary context. In W. R. Trevathan, E. O. Smith, and J. J. McKenna, eds., *Evolutionary Medicine,* 429–442. New York: Oxford University Press.

———. 2003. An evolutionary perspective on human physical activity: Implications for health. *Comparative Biochemistry and Physiology Part A* 136: 153–159.

Eaton, S. B., S. B. Eaton III, and L. Cordain. 2002. Evolution, diet, and health. In P. S. Ungar and M. F. Teaford, eds., *Human Diet: Its Origin and Evolution,* 7–17. Westport, Conn.: Bergin & Garvey.

Eaton, S. B., S. B. Eaton III, and M. J. Konner. 1997. Paleolithic nutrition revisited: A twelve-year retrospective on its nature and implications. *European Journal of Clinical Nutrition* 51: 207–216.

———. 1999. Paleolithic nutrition revisited. In W. R. Trevathan, E. O. Smith, and J. J. McKenna, eds., *Evolutionary Medicine,* 313–332. New York: Oxford University Press.

Eaton, S. B., and M. Konner. 1985. Paleolithic nutrition—A consideration of its nature and current implications. *New England Journal of Medicine* 312: 283–289.

Eaton, S. B., M. Konner, and M. Shostak. 1988. Stone agers in the fast lane: Chronic degenerative diseases in evolutionary perspective. *American Journal of Medicine* 84: 739–749.

Eaton, S. B., M. C. Pike, R. V. Short, N. C. Lee, J. Trussell, R. A. Hatcher, et al. 1994. Women's reproductive cancers in evolutionary context. *Quarterly Review of Biology* 69: 353–367.

Eaton, S. B., B. I. Strassman, R. M. Nesse, J. V. Neel, P. W. Ewald, G. C. Williams, et al. 2002. Evolutionary health promotion. *Preventive Medicine* 34: 109–118.

Eissa, M. K., M. S. Obhrai, M. F. Docker, S. S. Lynch, R. S. Sawers, and R. R. Newton. 1986. Follicular growth and endocrine profiles in spontaneous and induced conception cycles. *Fertility and Sterility* 45: 191–195.

Ekbom, A., and D. Trichopoulos. 1992. Evidence of prenatal influences on breast cancer risk. *Lancet* 340: 1015–1018.

Eliassen, A. H., S. A. Missmer, S. S. Tworoger, D. Spiegelman, R. L. Barbieri, M. Dowsett, and S. E. Hankinson. 2006. Endogenous steroid hormone concentrations and risk of breast cancer among premenopausal women. *Journal of the National Cancer Institute* 98: 1406–1415.

Ellison, P. T. 1982. Skeletal growth, fatness and menarcheal age: A comparison of two hypotheses. *Human Biology* 54: 269–281.

———. 1988. Human salivary steroids: Methodological issues and applications in physical anthropology. *Yearbook of Physical Anthropology* 31: 115–142.

———. 1990. Human ovarian function and reproductive ecology: New hypotheses. *American Anthropologist* 92: 933–952.

———. 1994. Salivary steroids and natural variation in human ovarian function. *Annals of the New York Academy of Sciences* 709: 287–298.

———. 1999. Reproductive ecology and reproductive cancers. In C. Panter-Brick and C. M. Worthman, eds., *Hormones, Health, and Behavior: A Socio-ecological and Lifespan Perspective,* 184–209. Cambridge: Cambridge University Press.

———. 2003a. Energetics and reproductive effort. *American Journal of Human Biology* 15: 342–351.

———. 2003b. *On Fertile Ground.* Cambridge, Mass.: Harvard University Press.

Ellison, P. T., R. G. Bribiescas, G. R. Bentley, B. C. Campbell, S. F. Lipson, C. Panter-Brick, and K. Hill. 2002. Population variation in age-related decline in male salivary testosterone. *Human Reproduction* 17: 3251–3253.

Ellison, P. T., and C. Lager. 1985. Exercise-induced menstrual disorders. *New England Journal of Medicine* 313: 825–826.

———. 1986. Moderate recreational running is associated with lowered salivary progesterone profiles in women. *American Journal of Obstetrics and Gynecology* 154: 1000–1003.

Ellison, P. T., S. F. Lipson, M. T. O'Rourke, G. R. Bentley, A. M. Harrigan, C. Panter-Brick, and V. J. Vitzthum. 1993. Population variation in ovarian function. *Lancet* 342: 433–434.

Ellison, P. T., N. R. Peacock, and C. Lager. 1986. Salivary progesterone and luteal function in two low-fertility populations of northeast Zaire. *Human Biology* 58: 473–483.

Eltis, D. 1982. Nutritional trends in Africa and the Americas—Heights of Africans, 1819–1839. *Journal of Interdisciplinary History* 12: 453–475.

Emanuel, I., C. Kimpo, and V. Moceri. 2004. The association of grandmaternal and maternal factors with maternal adult stature. *International Journal of Epidemiology* 33: 1243–1248.

Englyst, K. N., and H. N. Englyst. 2005. Carbohydrate bioavailability. *British Journal of Nutrition* 94: 1–11.

Eriksson, J. G., T. Forsen, J. Tuomilehto, P. D. Winter, C. Osmond, and D. J. P. Barker. 1999. Catch-up growth in childhood and death from coronary heart disease: Longitudinal study. *British Medical Journal* 318: 427–431.

ESHRE Capri Workshop Group. 2006. Nutrition and reproduction in women. *Human Reproduction Update* 12: 193–207.

Estioko-Griffin, A., and P. B. Griffin. 1981. Woman the hunter: The Agta. In F. Dahlberg, ed., *Woman the Gatherer,* 121–151. New Haven, Conn.: Yale University Press.

Evans, D. J., R. G. Hoffmann, R. K. Kalkhoff, and A. H. Kissebah. 1983. Relationship of androgenic activity to body-fat topography, fat-cell morphology, and metabolic aberrations in premenopausal women. *Journal of Clinical Endocrinology and Metabolism* 57: 304–310.

Fairweather-Tait, S., A. Prentice, K. G. Heumann, L. M. A. Jarjou, D. M. Stirling, S. G. Wharf, and J. R. Turnlund. 1995. Effect of calcium supplements and stage of lactation on the calcium-absorption efficiency of lactating women accustomed to low-calcium intakes. *American Journal of Clinical Nutrition* 62: 1188–1192.

Falsetti, L., E. Pasinetti, M. D. Mazzani, and A. Gastaldi. 1992. Weight-loss and menstrual cycle—clinical and endocrinologic evaluation. *Gynecological Endocrinology* 6: 49–56.

Farley, R. 1965. The demographic rates and social institutions of the 19th-century Negro population—A stable-population analysis. *Demography* 2: 386–398.

Feicht, C. B., T. S. Johnson, B. J. Martin, K. E. Sparks, and W. W. Wagner. 1978. Secondary amenorrhoea in athletes. *Lancet* 26: 1145–1146.

Feigelson, H. S., R. McKean-Cowdin, M. C. Pike, G. A. Coetzee, L. N. Kolonel, A. M. Y. Nomura, et al. 1999. Cytochrome P450c17 alpha gene (CYP17) polymorphism predicts use of hormone replacement therapy. *Cancer Research* 59: 3908–3910.

Feigelson, H. S., L. S. Shames, M. C. Pike, G. A. Coetzee, F. Z. Stanczyk, and B. E. Henderson. 1998. Cytochrome p450c17 alpha gene (CYP17) polymorphism is associated with serum estrogen and progesterone concentrations. *Cancer Research* 58: 585–587.

Ferenci, P. 2004. Review article: Diagnosis and current therapy of Wilson's disease. *Alimentary Pharmacology and Therapeutics* 19: 157–165.

Fessler, D. M. T. 2002. Reproductive immunosuppression and diet—An evolutionary perspective on pregnancy sickness and meat consumption. *Current Anthropology* 43: 19–61.

Festa-Bianchet, M. 1989. Individual differences, parasites, and the costs of reproduction for bighorn ewes (*Ovis canadensis*). *Journal of Animal Ecology* 58: 785–795.

Fiesch-Janys, D., T. Slanger, E. Mutschelknauss, S. Kropp, N. Obi, E. Vettorazzi, et al. 2008. Risk of different histological types of postmenopausal breast cancer by type and regimen of menopausal hormone therapy. *International Journal of Cancer* 123: 933–941.

Fildes, V., L. Marks, and H. Marland, eds. 1992. *Women and Children First: International Maternal and Infant Welfare 1870–1945*. London: Routledge.

Fillit, H. M. 2002. The role of hormone replacement therapy in the prevention of Alzheimer disease. *Archives of Internal Medicine* 162: 1934–1942.

Fleischman, D. S., and D. M. T. Fessler. 2011. Progesterone's effects on the psychology of disease avoidance: Support for the compensatory behavioral prophylaxis hypothesis. *Hormones and Behavior* 59: 271–275.

Fleming, A. F. 1994. Agriculture-related anemias. *British Journal of Biomedical Science* 51: 345–357.

Fogel, R. W., and S. L. Engerman. 1974. *Time on the Cross: The Economics of American Negro Slavery*. New York: W. W. Norton.

Follett, R. 2003. Heat, sex, and sugar: Pregnancy and childbearing in the slave quarters. *Journal of Family History* 28: 510–539.

Food and Agriculture Organization of the United Nations/World Health Organization/United Nations University (FAO/WHO/UNU). 1985. *Energy and Protein Requirements*. Technical Report Series 724. Geneva: World Health Organization.

Forsum, E., N. Kabir, A. Sadurskis, and K. Westerterp. 1992. Total energy expenditure of healthy Swedish women during pregnancy and lactation. *American Journal of Clinical Nutrition* 56: 334–342.

Francois, I., F. deZegher, C. Spiessens, T. Dhooghe, and D. Vanderschueren. 1997. Low birth weight and subsequent male subfertility. *Pediatric Research* 42: 899–901.

Fraser, G. E., and D. Shavlik. 1997. Risk factors, lifetime risk, and age at onset of breast cancer. *Annals of Epidemiology* 7: 375–382.

Frayling, T. M., N. J. Timpson, M. N. Weedon, E. Zeggini, R. M. Freathy, C. M. Lindgren, et al. 2007. A common variant in the FTO gene is associated with body mass index and predisposes to childhood and adult obesity. *Science* 316: 889–894.

Friedenreich, C. M., and A. E. Cust. 2008. Physical activity and breast cancer risk: Impact of timing, type and dose of activity and population subgroup effects. *British Journal of Sports Medicine* 42: 636–647.

Frisch, R. E. 1985. Fatness, menarche, and female fertility. *Perspectives in Biology and Medicine* 28: 611–633.

———. 1987. Body-fat, menarche, fitness and fertility. *Human Reproduction* 2: 521–533.

———. 1990. The right weight: Body fat, menarche and ovulation. *Clinical Obstetrics and Gynaecology* 4: 419–439.

Frisch, R. E., and J. W. McArthur. 1974. Menstrual cycles: Fatness as a determinant of minimum weight per height necessary for their maintenance or onset. *Science* 185: 949–951.

Frisch, R. E., G. Wyshak, J. Witschi, N. L. Albright, T. E. Albright, and I. Schiff. 1987. Lower lifetime occurrence of breast cancer and cancers of the reproductive system among former college athletes. *International Journal of Fertility* 32: 217–225.

Fuller, K. 2000. Lactose, rickets, and the coevolution of genes and culture. *Human Ecology* 28: 471–477.

Funahashi, H., M. Satake, S. Hasan, H. Sawai, R. A. Newman, H. A. Reber, et al. 2008. Opposing effects of n-6 and n-3 polyunsaturated fatty acids on pancreatic cancer growth. *Pancreas* 36: 353–362.

Furberg, A. S., G. Jasienska, N. Bjurstam, P. A. Torjesen, A. Emaus, S. F. Lipson, et al. 2005. Metabolic and hormonal profiles: HDL cholesterol as a plausible biomarker of breast cancer risk. The Norwegian EBBA Study. *Cancer Epidemiology, Biomarkers and Prevention* 14: 33–40.

Gajalakshmi, V. 2000. Diet and cancers of the stomach, breast and lung. *Asian Pacific Journal of Cancer Prevention* 1 (Suppl.): 39–43.

Galbarczyk, A. 2011. Unexpected changes in maternal breast size during pregnancy in relation to infant sex: An evolutionary interpretation. *American Journal of Human Biology* 23: 560–562.

Galletly, C., A. Clark, L. Tomlinson, and F. Blaney. 1996. Improved pregnancy rates for obese, infertile women following a group treatment program—An open pilot study. *General Hospital Psychiatry* 18: 192–195.

Gann, P. H., S. Giovanazzi, L. Van Horn, A. Branning, and R. T. Chatterton. 2001. Saliva as a medium for investigating intra- and interindividual differences in sex hormone levels in premenopausal women. *Cancer Epidemiology, Biomarkers and Prevention* 10: 59–64.

Gannage-Yared, M. H., R. Chemali, and N. Yaacoub. 2000. Hypovitaminosis D in a sunny country: Relation to lifestyle and bone markers. *Journal of Bone and Mineral Research* 15: 1856–1862.

Ganry, O. 2002. Phytoestrogen and breast cancer prevention. *European Journal of Cancer Prevention* 11: 519–522.

Garcia-Closas, M., J. Herbstman, M. Schiffman, A. Glass, and J. F. Dorgan. 2002. Relationship between serum hormone concentrations, reproductive history, alcohol consumption and genetic polymorphisms in pre-menopausal women. *International Journal of Cancer* 102: 172–178.

Garner, E. I. O., E. E. Stokes, R. S. Berkowitz, S. C. Mok, and D. W. Cramer. 2002. Polymorphisms of the estrogen-metabolizing genes CYP17 and catechol-O-methyltransferase and risk of epithelial ovarian cancer. *Cancer Research* 62: 3058–3062.

Garner, P., T. Smith, M. Baea, D. Lai, and P. Heywood. 1994. Maternal nutritional depletion in a rural area of Papua-New-Guinea. *Tropical and Geographical Medicine* 46: 169–171.

Garwood, E. R., A. S. Kumar, and V. Shim. 2008. Menopausal hormone therapy and breast cancer phenotype: Does dose matter? *Annals of Surgical Oncology* 15: 2526–2532.

Gavrilova, N. S., and L. A. Gavrilov. 2005. Human longevity and reproduction: An evolutionary perspective. In E. Voland, A. Chasiotis, and W. Schiefenhovel, eds., *Grandmotherhood: The Evolutionary Significance of the Second Half of Female Life,* 59–80. New Brunswick, N.J.: Rutgers University Press.

Gazzieri, D., M. Trevisani, F. Tarantini, P. Bechi, G. Masotti, G. F. Gensini, et al. 2006. Ethanol dilates coronary arteries and increases coronary flow via transient receptor potential vanilloid 1 and calcitonin gene-related peptide. *Cardiovascular Research* 70: 589–599.

Geelen, A., J. M. Schouten, C. Kamphuis, B. E. Stam, J. Burema, J. M. S. Renkema, et al. 2007. Fish consumption, n-3 fatty acids, and colorectal cancer: A meta-analysis of prospective cohort studies. *American Journal of Epidemiology* 166: 1116–1125.

Geggus, D. P. 1996. Slave and free colored women in Saint Domingue. In D. B. Gaspar and D. Clark Hine, eds., *More Than Chattel: Black Women and Slavery in the Americas,* 259–278. Bloomington: Indiana University Press.

George, D. S., P. M. Everson, J. C. Stevenson, and L. Tedrow. 2000. Birth intervals and early childhood mortality in a migrating Mennonite community. *American Journal of Human Biology* 12: 50–63.

Gerber, M. 1998. Fibre and breast cancer. *European Journal of Cancer Prevention* 7 (Suppl.): S63–S67.

Gettler, L. T., T. W. McDade, A. B. Feranil, and C. W. Kuzawa. 2011. Longitudinal evidence that fatherhood decreases testosterone in human males. *Proceedings of the National Academy of Sciences of the United States of America* 108: 16194–16199.

Gibbs, T., K. Cargill, L. S. Lieberman, and E. Reitz. 1980. Nutrition in a slave population: An anthropological examination. *Medical Anthropology* 4: 175–262.

Gibson, M. A., and R. Mace. 2006. An energy-saving development initiative increases birth rate and childhood malnutrition in rural Ethiopia. *PLoS Medicine* 3: 476–484.

Gluckman, P. D., and M. A. Hanson. 2005. *The Fetal Matrix: Evolution, Development and Disease.* Cambridge: Cambridge University Press.

Goldberg, G. R., A. M. Prentice, W. A. Coward, H. L. Davies, P. R. Murgatroyd, M. B. Sawyer, et al. 1991. Longitudinal assessment of the components of energy balance in well-nourished lactating women. *American Journal of Clinical Nutrition* 54: 788–798.

Goldenberg, R. L., S. P. Cliver, F. X. Mulvihill, C. A. Hickey, H. J. Hoffman, L. V. Klerman, and M. J. Johnson. 1996. Medical, psychosocial, and behavioral risk factors do not explain the increased risk for low birth weight among black women. *American Journal of Obstetrics and Gynecology* 175: 1317–1324.

Goldin, B. R., M. N. Woods, and D. L. Spiegelman. 1994. The effect of dietary fat and fiber on serum estrogen concentrations in premenopausal women under controlled dietary conditions. *Cancer* 74 (Suppl.): 1125–1131.

Goldstein, M. C., and C. M. Bell. 2002. Changing pattern of Tibetan nomadic pastoralism. In W. R. Leonard and M. H. Crawford, eds., *Human Biology of Pastoral Populations,* 131–150. Cambridge: Cambridge University Press.

Goodin, M. G., K. C. Fertuck, T. R. Zacharewski, and R. J. Rosengren. 2002. Estrogen receptor-mediated actions of polyphenolic catechins in vivo and in vitro. *Toxicological Sciences* 69: 354–361.

Gorai, I., K. Tanaka, M. Inada, H. Morinaga, Y. Uchiyama, R. Kikuchi, et al. 2003. Estrogen-metabolizing gene polymorphisms, but not estrogen receptor-alpha gene polymorphisms, are associated with the onset of menarche in healthy postmenopausal Japanese women. *Journal of Clinical Endocrinology and Metabolism* 88: 799–803.

Gosman, G. G., H. I. Katcher, and R. S. Legro. 2006. Obesity and the role of gut and adipose hormones in female reproduction. *Human Reproduction Update* 12: 585–601.

Gould, R. A. 1981. Comparative ecology of food-sharing in Australia and Northwest California. In R. S. O. Harding and G. Teleki, eds., *Omnivorous Primates,* 422–454. New York: Columbia University Press.

Graham, M. J., U. Larsen, and X. P. Xu. 1998. Son preference in Anhui Province, China. *International Family Planning Perspectives* 24: 72–77.

Grantham-McGregor, S. M., S. P. Walker, S. M. Chang, and C. A. Powell. 1997. Nutritional deficiencies and subsequent effects on mental and behavioral development in children. *Southeast Asian Journal of Tropical Medicine and Public Health* 28: 50–68.

Gray, P. B., and B. C. Campbell. 2009. Human males testosterone, pair-bonding, and fatherhood. In P. T. Ellison and P. B. Gray, eds., *Endocrinology of Social Relationship,* 270–293. Cambridge, Mass.: Harvard University Press.

Gray, P. B., S. M. Kahlenberg, E. S. Barrett, S. F. Lipson, and P. T. Ellison. 2002. Marriage and fatherhood are associated with lower testosterone in males. *Evolution and Human Behavior* 23: 193–201.

Greaves, M. F. 2001. *Cancer: The Evolutionary Legacy.* New York: Oxford University Press.

Green, A., V. Beral, and K. Moser. 1988. Mortality in women in relation to their childbearing history. *British Medical Journal* 297: 391–395.

Green, B. B., J. R. Daling, N. S. Weiss, J. M. Liff, and T. Koepsell. 1986. Exercise as a risk factor for infertility with ovulatory dysfunction. *American Journal of Public Health* 76: 432–436.

Green, B. B., N. S. Weiss, and J. R. Daling. 1988. Risk of ovulatory infertility in relation to body weight. *Fertility and Sterility* 50: 721–726.

Grether, W. F., and R. M. Yerkes. 1940. Weight norms and relations for chimpanzee. *American Journal of Physical Anthropology* 27: 181–197.

Grodstein, F., M. B. Goldman, and D. W. Cramer. 1994. Body-mass index and ovulatory infertility. *Epidemiology* 5: 247–250.

Gudmundsdottir, K., S. Thorlacius, J. G. Jonasson, B. F. Sigfusson, L. Tryggvadottir, and J. E. Eyfjord. 2003. CYP17 promoter polymorphism and breast cancer risk in males and females in relation to BRCA2 status. *British Journal of Cancer* 88: 933–936.

Guillermo-Tuazon, M. A., C. V. Barba, J. M. van Raaij, and J. G. Hautvast. 1992. Energy intake, energy expenditure, and body composition of poor rural Philippine women throughout the first 6 mo of lactation. *American Journal of Clinical Nutrition* 56: 874–80.

The Guinness Book of World Records 1998. 1998. New York: Bantam Books.

Gunnell, D. 2002. Commentary: Can adult anthropometry be used as a "biomarker" for prenatal and childhood exposures? *International Journal of Epidemiology* 31: 390–394.

Gussler, J. D. 1985. Commentary: Women's work and infant feeding in Oceania. In L. B. Marshall, ed., *Infant Care and Feeding in the South Pacific,* 319–329. New York: Gordon and Breach Science Publishers.

Gutman, H. G. 1976. *The Black Family in Slavery and Freedom, 1750–1925.* New York: Random House.

Gutman, H. G., and R. Sutch. 1976. Victorians all? The sexual mores and conduct of slaves and their masters. In P. A. David, ed., *Reckoning with Slavery: A Critical Study in the Quantitative History of American Negro Slavery,* 134–164. New York: Oxford University Press.

Haakstad, L. A. H., and K. Bo. 2011. Exercise in pregnant women and birth weight: A randomized controlled trial. *BMC Pregnancy and Childbirth* 11: 66.

Habuchi, T., L. Q. Zhang, T. Suzuki, R. Sasaki, N. Tsuchiya, H. Tachiki, et al. 2000. Increased risk of prostate cancer and benign prostatic hyperplasia associated with a CYP17 gene polymorphism with a gene dosage effect. *Cancer Research* 60: 5710–5713.

Haig, D. 1992. Genomic printing and the theory of parent-offspring conflict. *Seminars in Developmental Biology* 3: 153–160.

———. 2004. Evolutionary conflicts in pregnancy and calcium metabolism. *Placenta* 25 (Suppl. A): S10–S15.

———. 2008. Intimate relations: Evolutionary conflicts of pregnancy and childhood. In S. C. Stearns and J. C. Koella, eds., *Evolution in Health and Disease*, 65–76. New York: Oxford University Press.

Haiman, C. A., S. E. Hankinson, D. Spiegelman, G. A. Colditz, W. C. Willett, F. E. Speizer, et al. 1999. Relationship between a polymorphism in CYP17 with plasma hormone levels and breast cancer. *Cancer Research* 59: 1015–1020.

Haiman, C. A., M. J. Stampfer, E. Giovannucci, J. Ma, N. E. Decalo, P. W. Kantoff, and D. J. Hunter. 2001. The relationship between a polymorphism in CYP17 with plasma hormone levels and prostate cancer. *Cancer Epidemiology Biomarkers and Prevention* 10: 743–748.

Haines, C. J., S. M. Xing, K. H. Park, C. F. Holinka, and M. K. Ausmanas. 2005. Prevalence of menopausal symptoms in different ethnic groups of Asian women and responsiveness to therapy with three doses of conjugated estrogens/medroxyprogesterone acetate: The Pan-Asia Menopause (PAM) Study. *Maturitas* 52: 264–276.

Haines, M. R. 1985. Inequality and childhood mortality: A comparison of England and Wales, 1911, and the United States, 1900. *Journal of Economic History* 45: 885–912.

Hales, C. N., and D. J. P. Barker. 1992. Type-2 (non-insulin-dependent) diabetes mellitus—The thrifty phenotype hypothesis. *Diabetologia* 35: 595–601.

———. 2001. The thrifty phenotype hypothesis. *British Medical Bulletin* 60: 5–20.

Hallberg, L., L. Rossanderhulten, M. Brune, and A. Gleerup. 1992. Calcium and iron absorption—Mechanism of action and nutritional importance. *European Journal of Clinical Nutrition* 46: 317–327.

Hamajima, N., H. Iwata, Y. Obata, K. Matsuo, M. Mizutani, T. Iwase, et al. 2000. No association of the 5' promoter region polymorphism of CYP17 with breast cancer risk in Japan. *Japanese Journal of Cancer Research* 91: 880–885.

Hamilton, W. D. 1964. Genetical evolution of social behaviour, parts 1 and 2. *Journal of Theoretical Biology* 7: 1–52.

Hammond, K., and J. Diamond. 1994. Limits to dietary nutrient intake and intestinal nutrient uptake in lactating mice. *Physiological Zoology* 67: 282–303.

Handler, J. S., and R. S. Corruccini. 1986. Weaning among West Indian slaves: Historical and bioanthropological evidence from Barbados. *William and Mary Quarterly* 43: 111–117.

Hankinson, S. E., W. C. Willett, J. E. Manson, D. J. Hunter, G. A. Colditz, M. J. Stampfer, et al. 1995. Alcohol, height, and adiposity in relation to estrogen and prolactin levels in postmenopausal women. *Journal of the National Cancer Institute* 87: 1297–1302.

Hardman, A. E. 1999. Accumulation of physical activity for health gains: What is the evidence? *British Journal of Sports Medicine* 33: 87–92.

Harris, F. D. 1900. The supply of sterilized humanized milk for the use of infants in St. Helens. *British Medical Journal* 2: 427–431.

Hatun, S., O. Islam, F. Cizmecioglu, B. Kara, K. Babaoglu, F. Berk, and A. S. Gokalp. 2005. Subclinical vitamin D deficiency is increased in adolescent girls who wear concealing clothing. *Journal of Nutrition* 135: 218–222.

Hawkes, K., J. F. O'Connell, and N. G. B. Jones. 1997. Hadza women's time allocation, offspring provisioning, and the evolution of long postmenopausal life spans. *Current Anthropology* 38: 551–577.

Hawkes, K., J. F. O'Connell, N. G. B. Jones, H. Alvarez, and E. L. Charnov. 1998. Grandmothering, menopause, and the evolution of human life histories. *Proceedings of the National Academy of Sciences of the United States of America* 95: 1336–1339.

Hawkins, S. M., and M. M. Matzuk. 2008. The menstrual cycle—Basic biology. *Annals of the New York Academy of Sciences* 1135: 10–18.

Helle, S., V. Lummaa, and J. Jokela. 2002. Sons reduced maternal longevity in preindustrial humans. *Science* 296: 1085.

Henderson, B. E., and H. S. Feigelson. 2000. Hormonal carcinogenesis. *Carcinogenesis* 21: 427–433.

Henderson, B. E., R. K. Ross, and L. Bernstein. 1988. Estrogens as a cause of human cancer: The Richard and Hilda Rosenthal Foundation Award Lecture. *Cancer Research* 48: 246–253.

Heyward, V. H. 1991. *Advanced Fitness Assessments and Exercise Prescription.* Champaign, Ill.: Human Kinetics Publishers.

Hibbeln, J. R., L. R. G. Nieminen, T. L. Blasbalg, J. A. Riggs, and W. E. M. Lands. 2006. Healthy intakes of n-3 and n-6 fatty acids: Estimations considering worldwide diversity. *American Journal of Clinical Nutrition* 83 (Suppl.): 1483S–1493S.

Higman, B. W. 1979. Growth in Afro-Caribbean slave populations. *American Journal of Physical Anthropology* 50: 373–385.

———. 1984. *Slave Populations of the British Caribbean, 1807–1834.* Baltimore: Johns Hopkins University Press.

Hill, K, and A. M. Hurtado. 1996. *Ache Life History: The Ecology and Demography of a Foraging People.* Hawthorne, N.Y.: Aldine de Gruyter.

Hillard, P. J. A., and L. M. Nelson. 2003. Adolescent girls, the menstrual cycle, and bone health. *Journal of Pediatric Endocrinology and Metabolism* 16: 673–681.

Hinde, K. 2009. Richer milk for sons but more milk for daughters: Sex-biased investment during lactation varies with maternal life history in rhesus macaques. *American Journal of Human Biology* 21: 512–519.

Hinkula, M., A. Kauppila, S. Nayha, and E. Pukkala. 2006. Cause-specific mortality of grand multiparous women in Finland. *American Journal of Epidemiology* 163: 367–373.

Hinkula, M., E. Pukkala, P. Kyyronen, and A. Kauppila. 2001. Grand multiparity and the risk of breast cancer: Population-based study in Finland. *Cancer Causes and Control* 12: 491–500.

Hoff, C. 1999. Pregnancy, HLA allogeneic challenge, and implications for AIDS etiology. *Medical Hypotheses* 53: 63–68.

Holden, C., and R. Mace. 1997. Phylogenetic analysis of the evolution of lactose digestion in adults. *Human Biology* 69: 605–628.

Holzbeierlein, J. M., J. McIntosh, and J. B. Thrasher. 2005. The role of soy phytoestrogens in prostate cancer. *Current Opinion in Urology* 15: 17–22.

Homan, G. F., M. Davies, and R. Norman. 2007. The impact of lifestyle factors on reproductive performance in the general population and those undergoing infertility treatment: A review. *Human Reproduction Update* 13: 209–223.

Hong, C. C., H. J. Thompson, C. Jiang, G. L. Hammond, D. Tritchler, M. Yaffe, and N. F. Boyd. 2004. Association between the T27C polymorphism in the cytochrome P450 c17 alpha (CYP17) gene and risk factors for breast cancer. *Breast Cancer Research and Treatment* 88: 217–230.

Howe, J. C., W. V. Rumpler, and J. L. Seale. 1993. Energy expenditure by indirect calorimetry in premenopausal women: Variation within one menstrual cycle. *Journal of Nutritional Biochemistry* 4: 268–273.

Hrdy, S. B. 1999. *Mother Nature: A History of Mothers, Infants, and Natural Selection.* New York: Pantheon.

Huang, C. S., H. D. Chern, K. J. Chang, C. W. Cheng, S. M. Hsu, and C. Y. Shen. 1999. Breast cancer risk associated with genotype polymorphism of the estrogen-metabolizing genes CYP17, CYP1A1, and COMT: A multigenic study on cancer susceptibility. *Cancer Research* 59: 4870–4875.

Huang, Z., S. E. Hankinson, G. A. Colditz, M. J. Stampfer, D. J. Hunter, and J. E. Manson. 1997. Dual effects of weight and weight gain on breast cancer risk. *Journal of American Medical Association* 278: 1407–1411.

Hull, V. 1985. Breastfeeding, birth spacing, and social change in rural Java. In V. Hull and M. Simpson, eds., *Breastfeeding, Child Health and Child Spacing: Cross-cultural Perspectives,* 78–108. London: Croom Helm.

Humphries, K. H., I. C. D. Westendorp, M. L. Bots, J. J. Spinelli, R. G. Carere, A. Hofman, and J. C. M. Witteman. 2001. Parity and carotid artery atherosclerosis in elderly women—The Rotterdam Study. *Stroke* 32: 2259–2264.

Hunter, J. M. 1967. Seasonal hunger in a part of the West African Savanna: A survey of bodyweights in Nangodi, North-East Ghana. *Transactions of the Institute of British Geographers* 41: 167–185.

Huo, D., C. Adebamowo, T. Ogundiran, E. Akang, O. Campbell, A. Adenipekun, et al. 2008. Parity and breastfeeding are protective against breast cancer in Nigerian women. *British Journal of Cancer* 98: 992–996.

Hussain, M., M. Banerjee, F. H. Sarkar, Z. Djuric, M. N. Pollak, D. Doerge, et al. 2003. Soy isoflavones in the treatment of prostate cancer. *Nutrition and Cancer* 47: 111–117.

Huttly, S. R. A., C. G. Victora, F. C. Barros, and J. P. Vaughan. 1992. Birth spacing and child health in urban Brazilian children. *Pediatrics* 89: 1049–1054.

Hytten, F. E., and G. Chamberlain. 1991. *Clinical Physiology in Obstetrics,* 2nd ed. Oxford: Blackwell Scientific Publications.

Hytten, F. E., and I. Leitch. 1971. *The Physiology of Human Pregnancy,* 2nd ed. Oxford: Blackwell Scientific Publications.

Ibanez, L., N. Potau, G. Enriquez, and F. De Zegher. 2000. Reduced uterine and ovarian size in adolescent girls born small for gestational age. *Pediatric Research* 47: 575–577.

Ibanez, L., N. Potau, A. Ferrer, F. Rodriguez-Hierro, M. V. Marcos, and F. De Zegher. 2002. Reduced ovulation rate in adolescent girls born small for gestational age. *Journal of Clinical Endocrinology and Metabolism* 87: 3391–3393.

Idler, E. L., and Y. Benyamini. 1997. Self-rated health and mortality: A review of twenty-seven community studies. *Journal of Health and Social Behavior* 38: 21–37.

Ikeda, T., K. Makita, K. Ishitani, K. Takamatsu, F. Horiguchi, and S. Nozawa. 2005. Status of climacteric symptoms among middle-aged to elderly Japanese women: Comparison of general healthy women with women presenting at a menopausal clinic. *Journal of Obstetrics and Gynaecology Research* 31: 164–171.

Illingworth, P. J., R. T. Jung, P. W. Howie, P. Leslie, and T. E. Isles. 1986. Diminution in energy expenditure during lactation. *British Medical Journal* 292: 437–441.

Imhoff-Kunsch, B., R. Flores, O. Dary, and R. Martorell. 2007. Wheat flour fortification is unlikely to benefit the neediest in Guatemala. *Journal of Nutrition* 137: 1017–1022.

Immink, M. D. C. 1979. Impact of energy supplementation on daily energy intake and energy expenditure levels of Guatemalan sugarcane cutters. *Federation Proceedings* 38: 866.

Institute of Medicine. 1992. *Nutrition Issues in Developing Countries.* Washington, D.C.: National Academy Press.

Jablonka, E., and M. J. Lamb. 2005. *Evolution in Four Dimensions: Genetic, Epigenetic, Behavioral, and Symbolic Variation in the History of Life.* Cambridge, Mass.: MIT Press.

Jablonski, N. G., and G. Chaplin. 2010. Human skin pigmentation as an adaptation to UV radiation. *Proceedings of the National Academy of Sciences of the United States of America* 107: 8962–8968.

Jakicic, J. M., R. R. Wing, B. A. Butler, and R. J. Robertson. 1995. Prescribing exercise in multiple short bouts versus one continuous bout—Effects on adherence, cardiorespiratory fitness, and weight-loss in overweight women. *International Journal of Obesity* 19: 893–901.

Jamison, C. S., L. L. Cornell, P. L. Jamison, and H. Nakazato. 2002. Are all grandmothers equal? A review and a preliminary test of the "grandmother hypothesis" in Tokugawa Japan. *American Journal of Physical Anthropology* 119: 67–76.

Jarjou, L. M., A. Prentice, Y. Sawo, M. A. Laskey, J. Bennett, G. R. Goldberg, and T. J. Cole. 2006. Randomized, placebo-controlled, calcium supplementation study in pregnant Gambian women: Effects on breast-milk calcium concentrations and infant birth weight, growth, and bone mineral accretion in the first year of life. *American Journal of Clinical Nutrition* 83: 657–666.

Jasienska, G. 1996. Energy expenditure and ovarian function in Polish rural women. PhD diss., Harvard University.

———. 2001. Why energy expenditure causes reproductive suppression in women: An evolutionary and bioenergetic perspective. In P. T. Ellison, ed., *Reproductive Ecology and Human Evolution*, 59–85. New York: Aldine de Gruyter.

———. 2003. Energy metabolism and the evolution of reproductive suppression in the human female. *Acta Biotheoretica* 51: 1–18.

———. 2009. Low birth weight of contemporary African Americans: An intergenerational effect of slavery? *American Journal of Human Biology* 21: 16–24.

Jasienska, G., and P. T. Ellison. 1998. Physical work causes suppression of ovarian function in women. *Proceedings of the Royal Society of London B, Biological Sciences* 265: 1847–1851.

———. 2004. Energetic factors and seasonal changes in ovarian function in women from rural Poland. *American Journal of Human Biology* 16: 563–580.

Jasienska, G., and M. Jasienski. 2008. Interpopulation, interindividual, intercycle, and intracycle natural variation in progesterone levels: A quantitative assessment and implications for population studies. *American Journal of Human Biology* 20: 35–42.

Jasienska, G., M. Kapiszewska, P. T. Ellison, M. Kalemba-Drozdz, I. Nenko, I. Thune, and A. Ziomkiewicz. 2006. CYP17 genotypes differ in salivary 17-beta estradiol levels: A study based on hormonal profiles from entire menstrual cycles. *Cancer Epidemiology Biomarkers and Prevention* 15: 2131–2135.

Jasienska, G., I. Nenko, and M. Jasienski. 2006. Daughters increase longevity of fathers, but daughters and sons equally reduce longevity of mothers. *American Journal of Human Biology* 18: 422–425.

Jasienska, G., and I. Thune. 2001a. Lifestyle, hormones, and risk of breast cancer. *British Medical Journal* 322: 586–587.

———. 2001b. Lifestyle, progesterone and breast cancer. *British Medical Journal* 323: 1002.

Jasienska, G., I. Thune, and P. T. Ellison. 2000. Energetic factors, ovarian steroids and the risk of breast cancer. *European Journal of Cancer Prevention* 9: 231–239.

———. 2006. Fatness at birth predicts adult susceptibility to ovarian suppression: An empirical test of the Predictive Adaptive Response hypothesis. *Proceedings of the National Academy of Sciences of the United States of America* 103: 12759–12762.

Jasienska, G., A. Ziomkiewicz, P. T. Ellison, S. F. Lipson, and I. Thune. 2004. Large breasts and narrow waists indicate high reproductive potential in women. *Proceedings of the Royal Society of London B, Biological Sciences* 271: 1213–1217.

Jasienska, G., A. Ziomkiewicz, M. Gorkiewicz, and A. Pajak. 2005. Body mass, depressive symptoms and menopausal status: An examination of the "jolly fat" hypothesis. *Women's Health Issues* 15: 145–151.

Jasienska, G., A. Ziomkiewicz, S. F. Lipson, I. Thune, and P. T. Ellison. 2006. High ponderal index at birth predicts high estradiol levels in adult women. *American Journal of Human Biology* 18: 133–140.

Jasienska, G., A. Ziomkiewicz, I. Thune, S. F. Lipson, and P. T. Ellison. 2006. Habitual physical activity and estradiol levels in women of reproductive age. *European Journal of Cancer Prevention* 15: 439–445.

Jernstrom, H., J. Lubinski, H. T. Lynch, P. Ghadirian, S. Neuhausen, C. Isaacs, et al. 2004. Breast-feeding and the risk of breast cancer in BRCA1 and BRCA2 mutation carriers. *Journal of the National Cancer Institute* 96: 1094–1098.

Johnson, W. G., S. A. Corrigan, C. R. Lemmon, K. B. Bergeron, and A. H. Crusco. 1994. Energy regulation over the menstrual cycle. *Physiology and Behavior* 56: 523–527.

Johnston, E., S. Johnston, P. McLeod, and M. Johnston. 2004. The relation of body mass index to depressive symptoms. *Canadian Journal of Public Health* 95: 179–183.

Joyner, C. W. 1971. Life under the "peculiar institution"—Selections from the Slave Narrative Collection. *Journal of American Folklore* 84: 453–455.

Judd, J. T., B. A. Clevidence, R. A. Muesing, J. Wittes, M. E. Sunkin, and J. J. Podczasy. 1994. Dietary trans fatty acids: Effects on plasma lipids and lipoproteins of healthy men and women. *American Journal of Clinical Nutrition* 59: 861–868.

Kado, N., J. Kitawaki, H. Obayashi, H. Ishihara, H. Koshiba, I. Kusuki, et al. 2002. Association of the CYP17 gene and CYP19 gene polymorphisms with risk of endometriosis in Japanese women. *Human Reproduction* 17: 897–902.

Kaneda, N., C. Nagata, M. Kabuto, and H. Shimizu. 1997. Fat and fiber intakes in relation to serum estrogen concentration in premenopausal Japanese women. *Nutrition and Cancer* 27: 279–283.

Kao, Y. C., C. B. Zhou, M. Sherman, C. A. Laughton, and S. Chen. 1998. Molecular basis of the inhibition of human aromatase (estrogen synthetase) by flavone and isoflavone phytoestrogens: A site-directed mutagenesis study. *Environmental Health Perspectives* 106: 85–92.

Kapiszewska, M., M. Miskiewicz, P. T. Ellison, I. Thune, and G. Jasienska. 2006. High tea consumption diminishes salivary 17-beta estradiol concentration in Polish women. *British Journal of Nutrition* 95: 989–995.

Kaplan, H. 1997. The evolution of the human life course. In K. Wachter and C. Finch, eds., *Between Zeus and the Salmon: The Biodemography of Longevity,* 175–211. Washington, D.C.: National Academy Press.

Kaput, J. 2004. Diet-disease gene interactions. *Nutrition* 20: 26–31.

Karlsson, C., K. J. Obrant, and M. Karlsson. 2001. Pregnancy and lactation confer reversible bone loss in humans. *Osteoporosis International* 12: 828–834.

Katz, D. F., D. A. Slade, and S. T. Nakajima. 1997. Analysis of pre-ovulatory changes in cervical mucus hydration and sperm penetrability. *Advances in Contraception* 13: 143–151.

Katz, M. M. 1985. Infant care in a group of outer Fiji Islands. In L. B. Marshall, ed., *Infant Care and Feeding in the South Pacific,* 269–292. New York: Gordon and Breach Science Publishers.

Katzmarzyk, P. T., W. R. Leonard, M. H. Crawford, and R. I. Sukerinik. 1994. Resting metabolic rate and daily energy expenditure among two indigenous Siberian populations. *American Journal of Human Biology* 6: 719–730.

Kaufert, P., M. Lock, S. McKinlay, Y. Beyenne, J. Coope, D. Davis, et al. 1986. Menopause research—The Korpilampi Workshop. *Social Science and Medicine* 22: 1285–1289.

Keen, A. D., and B. L. Drinkwater. 1997. Irreversible bone loss in former amenorrheic athletes. *Osteoporosis International* 7: 311–315.

Kerstetter, J. E., K. O. O'Brien, and K. L. Insogna. 2003. Dietary protein, calcium metabolism, and skeletal homeostasis revisited. *American Journal of Clinical Nutrition* 78 (Suppl.): 584S–592S.

Key, T. J. A., J. Chen, D. Y. Wang, M. C. Pike, and J. Boreharn. 1990. Sex hormones in women in rural China and in Britain. *British Journal of Cancer* 62: 631–636.

Key, T. J. A., and M. C. Pike. 1988. The role of oestrogens and progestagens in the epidemiology and prevention of breast cancer. *European Journal of Clinical Oncology* 24: 29–43.

Khan, A. D., D. G. Schroeder, R. Martorell, J. D. Haas, and J. Rivera. 1996. Early childhood determinants of age at menarche in rural Guatemala. *American Journal of Human Biology* 8: 717–723.

Khan, K. S., P. F. W. Chien, and N. B. Khan. 1998. Nutritional stress of reproduction—A cohort study over two consecutive pregnancies. *Acta Obstetricia et Gynecologica Scandinavica* 77: 395–401.

King, J. C., N. F. Butte, M. N. Bronstein, L. E. Koop, and S. A. Lindquist. 1994. Energy metabolism during pregnancy: Influence of maternal energy status. *American Journal of Clinical Nutrition* 59: S439–S445.

King, M. C., J. H. Marks, and J. B. Mandell. 2003. Breast and ovarian cancer risks due to inherited mutations in BRCA1 and BRCA2. *Science* 302: 643–646.

King, W. 1995. *Stolen Childhood: Slave Youth in Nineteenth-Century America.* Bloomington: Indiana University Press.

Kington, R., L. Lillard, and J. Rogowski. 1997. Reproductive history, socioeconomic status, and self-reported health status of women aged 50 years or older. *American Journal of Public Health* 87: 33–37.

Kiple, K. F. 1984. *The Caribbean Slave: A Biological History.* Cambridge: Cambridge University Press.

Kiple, K. F., and W. Himmelsteib King. 1981. *Another Dimension to the Black Diaspora: Diet, Disease, and Racism.* Cambridge: Cambridge University Press.

Kiple, K. F., and V. H. Kiple. 1977a. Black tongue and Black men—Pellagra and slavery in the Antebellum South. *Journal of Southern History* 43: 411–428.

———. 1977b. Slave child mortality—Some nutritional answers to a perennial puzzle. *Journal of Social History* 10: 284–309.

Kirchengast, S. 1994. Interaction between sex-hormone levels and body dimensions in postmenopausal women. *Human Biology* 66: 481–494.

———. 2000. Differential reproductive success and body size in !Kung San people from northern Namibia. *Collegium Antropologicum* 24: 121–132.

Kirchengast, S., and S. Hartmann. 1998. Maternal prepregnancy weight status and pregnancy weight gain as major determinants for newborn weight and size. *Annals of Human Biology* 25: 17–28.

Kirkwood, T. B. L., and R. Holliday. 1979. Evolution of aging and longevity. *Proceedings of the Royal Society of London B, Biological Sciences* 205: 531–546.

Kirkwood, T. B. L., P. Kapahi, and D. P. Shanley. 2000. Evolution, stress, and longevity. *Journal of Anatomy* 197: 587–590.

Kittles, R. A., R. K. Panguluri, W. D. Chen, A. Massac, C. Ahaghotu, A. Jackson, et al. 2001. CYP17 promoter variant associated with prostate cancer aggressiveness in African Americans. *Cancer Epidemiology Biomarkers and Prevention* 10: 943–947.

Knight, D. C., and J. A. Eden. 1995. Phytoestrogens—A short review. *Maturitas* 22: 167–175.

Kochar, D. K., I. Thanvi, A. Joshi, Z. Z. Subhakaran, S. Aseri, and B. L. Kumawat. 1998. Falciparum malaria and pregnancy. *Indian Journal of Malariology* 35: 123–130.

Komlos, J. 1994. The height of runaway slaves in Colonial America, 1720–1770. In J. Komlos, ed., *Stature, Living Standards, and Economic Development,* 93–116. Chicago: University of Chicago Press.

Konarzewski, M., and J. Diamond. 1994. Peak sustained metabolic rate and its individual variation in cold-stressed mice. *Physiological Zoology* 67: 1186–1212.

Konner, M., and C. Worthman. 1980. Nursing frequency, gonadal function, and birth-spacing among !Kung hunter-gatherers. *Science* 207: 788–791.

Kopp-Hoolihan, L. E., M. D. van Loan, W. W. Wong, and J. C. King. 1999. Longitudinal assessment of energy balance in well-nourished, pregnant women. *American Journal of Clinical Nutrition* 69: 697–704.

Kovacs, C. S. 2005. Calcium and bone metabolism during pregnancy and lactation. *Journal of Mammary Gland Biology and Neoplasia* 10: 105–118.

Kramer, M. S. 1996. Nutritional advice in pregnancy. *Cochrane Database of Systematic Reviews* 4: CD000149.

Kristensen, V. N., E. K. Haraldsen, K. B. Anderson, P. E. Lonning, B. Erikstein, R. Karesen, et al. 1999. CYP17 and breast cancer risk: The polymorphism in the 5' flanking area of the gene does not influence binding to Sp-1. *Cancer Research* 59: 2825–2828.

Kritz-Silverstein, D., E. Barrett-Connor, and D. L. Wingard. 1989. The effect of parity on the later development of non-insulin-dependent diabetes-mellitus or impaired glucose-tolerance. *New England Journal of Medicine* 321: 1214–1219.

Kritz-Silverstein, D., E. Barrett-Connor, D. L. Wingard, and N. J. Friedlander. 1994. Relation of pregnancy history to insulin levels in older, nondiabetic women. *American Journal of Epidemiology* 140: 375–382.

Krzyk, J., and D. Steinhagen. 2007. Decyzja Anny Radosz. *Wysokie Obcasy,* 10–16.

Kuiper, G. J. M., B. Carlsson, K. Grandien, E. Enmark, J. Haggblad, S. Nilsson, and J. A. Gustafsson. 1997. Comparison of the ligand binding specificity and transcript tissue distribution of estrogen receptors alpha and beta. *Endocrinology* 138: 863–870.

Kuligina, E. S., A. V. Togo, E. N. Suspitsin, M. Y. Grigoriev, K. M. Pozharisskiy, O. L. Chagunava, et al. 2000. CYP17 polymorphism in the groups of distinct breast cancer susceptibility: Comparison of patients with the bilateral disease vs. monolateral breast cancer patients vs. middle-aged female controls vs. elderly tumor-free women. *Cancer Letters* 156: 45–50.

Kumar, N. B., A. Cantor, K. Allen, D. Riccardi, K. Besterman-Dahan, J. Seigne, et al. 2004. The specific role of isoflavones in reducing prostate cancer risk. *Prostate* 59: 141–147.

Kumar, N. B., A. Cantor, K. Allen, D. Riccardi, and C. E. Cox. 2002. The specific role of isoflavones on estrogen metabolism in premenopausal women. *Cancer* 94: 1166–1174.

Kusin, J. A., S. Kardjati, J. M. Houtkooper, and U. H. Renqvist. 1992. Energy supplementation during pregnancy and postnatal growth. *Lancet* 340: 623–626.

Kuzawa, C. W. 1998. Adipose tissue in human infancy and childhood: An evolutionary perspective. *Yearbook of Physical Anthropology:* 177–209.

———. 2004. Modeling fetal adaptation to nutrient restriction: Testing the fetal origins hypothesis with a supply-demand model. *Journal of Nutrition* 134: 194–200.

———. 2005. Fetal origins of developmental plasticity: Are fetal cues reliable predictors of future nutritional environments? *American Journal of Human Biology* 17: 5–21.

———. 2008. The developmental origins of adult health: Intergenerational inertia in adaptation and disease. In W. R. Trevathan, E. O. Smith, and J. J. McKenna, eds., *Evolutionary Medicine and Health: New Perspectives,* 325–349. New York: Oxford University Press.

Kvale, G. 1992. Reproductive factors in breast cancer epidemiology. *Acta Oncologica* 31: 187–194.

Lager, C., and P. T. Ellison. 1990. Effect of moderate weight loss on ovarian function assessed by salivary progesterone measurements. *American Journal of Human Biology* 2: 303–312.

Lai, J., D. Vesprini, W. Chu, H. Jernstrom, and S. A. Narod. 2001. CYP gene polymorphisms and early menarche. *Molecular Genetics and Metabolism* 74: 449–457.

Lake, J. K., C. Power, and T. J. Cole. 1997. Women's reproductive health: The role of body mass index in early and adult life. *International Journal of Obesity* 21: 432–438.

Lamb, J. D., A. M. Zamah, S. H. Shen, C. McCulloch, M. I. Cedars, and M. P. Rosen. 2011. Follicular fluid steroid hormone levels are associated with fertilization outcome after intracytoplasmic sperm injection. *Obstetrical and Gynecological Survey* 66: 218–220.

Lancaster, J. B. 1986. Human adolescence and reproduction: An evolutionary perspective. In J. B. Lancaster and B. A. Hamburg, eds., *School-Age Pregnancy and Parenthood,* 17–38. New York: Aldine de Gruyter.

Lands, W. E. M. 2005. Dietary fat and health: The evidence and the politics of prevention: Careful use of dietary fats can improve life and prevent disease. *Annals of the New York Academy of Sciences* 1055: 179–192.

Larsen, C. S. 1984. Health and disease in prehistoric Georgia: The transition to agriculture. In M. N. Cohen and G. J. Armelagos, eds., *Paleopathology at the Origins of Agriculture*, 367–392. Orlando, Fla.: Academic Press.

——. 1995. Biological changes in human populations with agriculture. *Annual Review of Anthropology* 24: 185–213.

——. 2002. Post-Pleistocene human evolution: Bioarcheology of the agricultural transition. In P. S. Ungar and M. F. Teaford, eds., *Human Diet: Its Origin and Evolution,* 19–35. Westport, Conn.: Bergin & Garvey.

Laskey, M. A., and A. Prentice. 1997. Effect of pregnancy on recovery of lactational bone loss. *Lancet* 349: 1518–1519.

Lasley, B. L., N. Santoro, J. F. Randolf, E. B. Gold, S. Crawford, G. Weiss, et al. 2002. The relationship of circulating dehydroepiandrosterone, testosterone, and estradiol to stages of the menopausal transition and ethnicity. *Journal of Clinical Endocrinology and Metabolism* 87: 3760–3767.

Law, M., and N. Wald. 1999. Why heart disease mortality is low in France: The time lag explanation. *British Medical Journal* 318: 1471–1476.

Lawlor, D. A., J. R. Emberson, S. Ebrahim, P. H. Whincup, S. G. Wannamethee, M. Walker, and G. D. Smith. 2003. Is the association between parity and coronary heart disease due to biological effects of pregnancy or adverse lifestyle risk factors associated with child-rearing?—Findings from the British women's heart and health study and the British regional heart study. *Circulation* 107: 1260–1264.

Lawrence, M., W. A. Coward, F. Lawrence, T. J. Cole, and R. G. Whitehead. 1987. Fat gain during pregnancy in rural African women: The effect of season and dietary status. *American Journal of Clinical Nutrition* 45: 1442–1450.

Lawrence, M., F. Lawrence, W. A. Coward, T. J. Cole, and R. G. Whitehead. 1987. Energy requirements of pregnancy in the Gambia. *Lancet* 2: 1072–1076.

Lawrence, M., and R. G. Whitehead. 1988. Physical activity and total energy expenditure of child-bearing Gambian village women. *European Journal of Clinical Nutrition* 42: 145–160.

Le Bourg, E., B. Thon, J. Legare, B. Desjardins, and H. Charbonneau. 1993. Reproductive life of French-Canadians in the 17–18th-centuries—A search for a trade-off between early fecundity and longevity. *Experimental Gerontology* 28: 217–232.

Lechtig, A., R. Martorell, H. Delgado, C. Yarbrough, and R. E. Klein. 1978. Food supplementation during pregnancy, maternal anthropometry and birth weight in a Guatemalan rural population. *Journal of Tropical Pediatrics and Environmental Child Health* 24: 217–222.

Lechtig, A., C. Yarbrough, H. Delgado, J. P. Habicht, R. Martorell, and R. E. Klein. 1975. Influence of maternal nutrition on birth weight. *American Journal of Clinical Nutrition* 28: 1223–1233.

Lee, I. M., and R. S. Paffenbarger. 2000. Associations of light, moderate, and vigorous intensity physical activity with longevity—The Harvard Alumni Health Study. *American Journal of Epidemiology* 151: 293–299.

Lee, I. M., K. M. Rexrode, N. R. Cook, C. H. Hennekens, and J. E. Buring. 2001. Physical activity and breast cancer risk: The Women's Health Study (United States). *Cancer Causes and Control* 12: 137–145.

Lee, R. B. 1968. What hunters do for a living, or how to make out on scarce resources. In R. B. Lee and I. DeVore, eds., *Man the Hunter,* 30–48. Chicago: Aldine.

———. 1979. *The !Kung San: Men, Women, and Work in a Foraging Society.* Cambridge: Cambridge University Press.

———. 1984. *The Dobe !Kung.* New York: Holt, Rinehart and Winston.

Lenton, E. A., G. F. Lawrence, R. A. Coleman, and I. D. Cooke. 1983. Individual variation in gonadotrophin and steroid concentration and in lengths of the follicular and luteal phases in women with regular menstrual cycles. *Clinical Reproduction and Fertility* 2: 143–150.

Leonard, W. R. 2008. Lifestyle, diet, and disease: Comparative perspectives on the determinants of chronic health risks. In S. C. Stearns and J. C. Koella, eds., *Evolution in Health and Disease,* 265–276. New York: Oxford University Press.

Leonard, W. R., V. A. Galloway, E. Ivakine, L. Osipova, and M. Kazakovtseva. 2002. Ecology, health and lifestyle change among the Evenki herders of Siberia. In W. R. Leonard and M. H. Crawford, eds., *Human Biology of Pastoral Populations,* 206–235. Cambridge: Cambridge University Press.

Leonard, W. R., and M. L. Robertson. 1992. Nutritional requirements and human evolution: A bioenergetic model. *American Journal of Human Biology* 4: 179–195.

Leopold, A. S., M. Erwin, J. Oh, and B. Browning. 1976. Phytoestrogens—Adverse effects on reproduction in California quail. *Science* 191: 98–100.

Lesthaeghe, R., D. Meekers, and G. Kaufmann. 1994. Postpartum abstinence, polygyny, and age at marriage: A macro-level analysis of sub-Saharan societies. In C. Bledsoe and G. Pison, eds., *Nuptiality in Sub-Saharan Africa: Contemporary Anthropological and Demographic Perspectives,* 25–54. Oxford: Clarendon Press.

Lewontin, R. C. 1992. *Biology as Ideology: The Doctrine of DNA.* New York: HarperPerennial.

———. 2000. *The Triple Helix.* Cambridge, Mass.: Harvard University Press.

Li, H., H. X. Barnhart, A. D. Stein, and R. Martorell. 2003. Effects of early childhood supplementation on the educational achievement of women. *Pediatrics* 112: 1156–1162.

Lichtenstein, A. H., L. J. Appel, M. Brands, M. Carnethon, S. Daniels, H. A. Franch, et al. 2006. Diet and lifestyle recommendations revision 2006—A scientific statement from the American Heart Association Nutrition Committee. *Circulation* 114: 82–96.

Lieberman, E. H., M. D. Gerhard, A. Uehata, B. W. Walsh, A. P. Selwyn, P. Ganz, et al. 1994. Estrogen improves endothelium-dependent, flow-mediated vasodilation in postmenopausal women. *Annals of Internal Medicine* 121: 936–941.

Lips, P. 2007. Vitamin D status and nutrition in Europe and Asia. *Journal of Steroid Biochemistry and Molecular Biology* 103: 620–625.

Lipson, S. F. 2001. Metabolism, maturation, and ovarian function. In P. T. Ellison, ed., *Reproductive Ecology and Human Evolution*, 235–248. New York: Aldine de Gruyter.

Lipson, S. F., and P. T. Ellison. 1996. Comparison of salivary steroid profiles in naturally occurring conception and non-conception cycles. *Human Reproduction* 11: 2090–2096.

Little, M. A. 2002. Human biology, health, and ecology of nomadic Turkana pastoralists. In W. R. Leonard and M. H. Crawford, eds., *Human Biology of Pastoral Populations*, 151–182. Cambridge: Cambridge University Press.

Little, M. A., P. W. Leslie, and K. L. Campbell. 1992. Energy reserves and parity of nomadic and settled Turkana women. *American Journal of Human Biology* 4: 729–738.

Lo, S. F., C. M. Huang, H. C. Lin, C. H. Tsai, and F. J. Tsai. 2005. Association of CYP17 gene polymorphism and rheumatoid arthritis in Chinese patients in central Taiwan. *Rheumatology International* 25: 580–584.

Lock, M. 1993. *Encounters with Aging: Mythologies of Menopause in Japan and North America*. Berkeley: University of California Press.

Love, R. R., and J. Philips. 2002. Oophorectomy for breast cancer: History revisited. *Journal of the National Cancer Institute* 94: 1433–1434.

Lu, L. J. W., K. E. Anderson, J. J. Grady, F. Kohen, and M. Nagamani. 2000. Decreased ovarian hormones during a soya diet: Implications for breast cancer prevention. *Cancer Research* 60: 4112–4121.

Lu, Y. C., G. R. Bentley, P. H. Gann, K. R. Hodges, and R. T. Chatterton. 1999. Salivary estradiol and progesterone levels in conception and nonconception cycles in women: Evaluation of a new assay for salivary estradiol. *Fertility and Sterility* 71: 863–868.

Lund, E., E. Arnesen, and J. K. Borgan. 1990. Pattern of childbearing and mortality in married women—A national prospective study from Norway. *Journal of Epidemiology and Community Health* 44: 237–240.

Lund, T. D., D. J. Munson, M. E. Haldy, K. D. R. Setchell, E. D. Lephart, and R. J. Handa. 2004. Equol is a novel anti-androgen that inhibits prostate growth and hormone feedback. *Biology of Reproduction* 70: 1188–1195.

Lunn, P. G. 1994. Lactation and other metabolic loads affecting human reproduction. *Annals of the New York Academy of Sciences* 709: 77–85.

Lunn, P. G., S. Austin, and R. G. Whitehead. 1984. The effect of improved nutrition on plasma prolactin concentrations and postpartum infertility in lactating Gambian women. *American Journal of Clinical Nutrition* 39: 227–235.

Lunn, R. M., D. A. Bell, J. L. Mohler, and J. A. Taylor. 1999. Prostate cancer risk and polymorphism in 17 hydroxylase (CYP17) and steroid reductase (SRD5A2). *Carcinogenesis* 20: 1727–1731.

Lycett, J. E., R. I. M. Dunbar, and E. Voland. 2000. Longevity and the costs of reproduction in a historical human population. *Proceedings of the Royal Society of London B, Biological Sciences* 267: 31–35.

Mabilia, M. 2005. *Breast Feeding and Sexuality : Behaviour, Beliefs and Taboos among the Gogo Mothers in Tanzania*. New York: Berghahn Books.

Macfarlane, D. J., L. H. Taylor, and T. F. Cuddihy. 2006. Very short intermittent vs continuous bouts of activity in sedentary adults. *Preventive Medicine* 43: 332–336.

Mackie, G. 1996. Ending footbinding and infibulation: A convention account. *American Sociological Review* 61: 999–1017.

MacMahon, B. 2006. Epidemiology and the causes of breast cancer. *International Journal of Cancer* 118: 2373–2378.

Madhavapeddi, R., and B. S. Rao. 1992. Energy balance in lactating undernourished Indian women. *European Journal of Clinical Nutrition* 46: 349–354.

Magnus, P., L. S. Bakketeig, and R. Skjaerven. 1993. Correlations of birth weight and gestational age across generations. *Annals of Human Biology* 20: 231–238.

Maher, V. 1992. Breast-feeding in cross-cultural perspective: Paradoxes and proposals. In V. Maher, ed., *The Anthropology of Breast-Feeding: Natural Law or Social Construct,* 1–36. Oxford: St. Martin's Press.

Maklakov, A. A., S. J. Simpson, F. Zajitschek, M. D. Hall, J. Dessmann, F. Clissold, et al. 2008. Sex-specific fitness effects of nutrient intake on reproduction and lifespan. *Current Biology* 18: 1062–1066.

Malcolm, L. W. G. 1925. Note of the seclusion of girls among the Efik at Old Calabar. *Men* 25: 113–114.

Manor, O., Z. Eisenbach, A. Israeli, and Y. Friedlander. 2000. Mortality differentials among women: The Israel Longitudinal Mortality Study. *Social Science and Medicine* 51: 1175–1188.

Manson, J. E., J. Hsia, K. C. Johnson, J. E. Rossouw, A. R. Assaf, N. L. Lasser, et al. 2003. Estrogen plus progestin and the risk of coronary heart disease. *New England Journal of Medicine* 349: 523–534.

Margetts, B. M., S. M. Yusof, Z. Al Dallal, and A. A. Jackson. 2002. Persistence of lower birth weight in second generation South Asian babies born in the United Kingdom. *Journal of Epidemiology and Community Health* 56: 684–687.

Margo, R. A., and R. H. Steckel. 1983. Heights of native-born whites during the antebellum period. *Journal of Economic History* 43: 167–174.

Marks, L. 1992. Mothers, babies and hospitals: "The London" and the provision of maternity care in East London, 1870–1939. In V. Fildes, L. Marks, and H. Marland, eds., *Women and Children First: International Maternal and Infant Welfare 1870–1945,* 48–73. London: Routledge.

Marlowe, F. W. 2007. Hunting and gathering—The human sexual division of foraging labor. *Cross-Cultural Research* 41: 170–195.

Martin, R. M., G. D. Smith, S. Frankel, and D. Gunnell. 2004. Parents' growth in childhood and the birth weight of their offspring. *Epidemiology* 15: 308–316.

Martorell, R. 1996. The role of nutrition in economic development. *Nutrition Reviews* 54: S66–S71.

Maskarinec, G., A. A. Franke, A. E. Williams, S. Hebshi, C. Oshiro, S. Murphy, and F. Z. Stanczyk. 2004. Effects of a 2-year randomized soy intervention on sex hormone levels in premenopausal women. *Cancer Epidemiology Biomarkers and Prevention* 13: 1736–1744.

Massara, E. B. 1989. *Que Gordita! A Study of Weight among Women in a Puerto Rican Community.* New York: AMS Press.

McArdle, P. F., T. I. Pollin, J. R. O'Connell, J. D. Sorkin, R. Agarwala, A. A. Schaffer, et al. 2006. Does having children extend life span? A genealogical study of parity and longevity in the Amish. *Journals of Gerontology Series A: Biological Sciences and Medical Sciences* 61: 190–5.

McArdle, W. D., F. I. Katch, and V. L. Katch. 1986. *Exercise Physiology: Energy, Nutrition, and Human Performance.* Philadelphia: Lea & Febiger.

McAuley, P. A., J. N. Myers, J. P. Abella, S. Y. Tan, and V. F. Froelicher. 2007. Exercise capacity and body mass as predictors of mortality among male veterans with type 2 diabetes. *Diabetes Care* 30: 1539–1543.

McCann, S. E., K. B. Moysich, J. L. Freudenheim, C. B. Ambrosone, and P. G. Shields. 2002. The risk of breast cancer associated with dietary lignans differs by CYP17 genotype in women. *Journal of Nutrition* 132: 3036–3041.

McCargar, L. J., D. Simmons, N. Craton, J. E. Taunton, and C. L. Birmingham. 1993. Physiological effects of weight cycling in female lightweight rowers. *Canadian Journal of Applied Physiology* 18: 291–303.

McCormick, M. C. 1985. The contribution of low birth-weight to infant-mortality and childhood morbidity. *New England Journal of Medicine* 312: 82–90.

McDade, T. W. 2005. The ecologies of human immune function. *Annual Review of Anthropology* 34: 495–521.

McKinlay, S. M., and J. B. McKinlay. 1986. Aging in a healthy population. *Social Science and Medicine* 23: 531–535.

McNamara, J. P. 1995. Role and regulation of metabolism in adipose tissue during lactation. *Journal of Nutritional Biochemistry* 6: 120–129.

McNeely, M. J., and M. R. Soules. 1988. The diagnosis of luteal phase deficiency: A critical review. *Fertility and Sterility* 50: 1–9.

McNeilly, A. S., C. C. K. Tay, and A. Glasier. 1994. Physiological mechanisms underlying lactational amenorrhea. In J. W. Wood, ed., *Human Reproductive Ecology: Interactions of Environment, Fertility, and Behavior*, 145–155. New York: New York Academy of Sciences.

McTiernan, A. 2008. Mechanisms linking physical activity with cancer. *Nature Reviews Cancer* 8: 205–211.

Meijer, G. A. L., K. R. Westerterp, W. H. M. Saris, and F. ten Hoor. 1992. Sleeping metabolic rate in relation to body composition and the menstrual cycle. *American Journal of Clinical Nutrition* 55: 637–640.

Merchant, K. S., and R. Martorell. 1988. Frequent reproductive cycling: Does it lead to nutritional depletion of mothers? *Progress in Food and Nutrition Science* 12: 339–369.

Mettlin, C. 1999. Global breast cancer mortality statistics. *CA: A Cancer Journal for Clinicians* 49: 138–144.

Michels, K. B., A. R. Mohllajee, E. Roset-Bahmanyar, G. P. Beehler, and K. B. Moysich. 2007. Diet and breast cancer—A review of the prospective observational studies. *Cancer* 109: 2712–2749.

Michels, K. B., D. Trichopoulos, J. M. Robins, B. A. Rosner, J. E. Manson, D. J. Hunter, et al. 1996. Birthweight as a risk factor for breast cancer. *Lancet* 348: 1542–1546.

Miklos, G. L. G. 2005. The Human Cancer Genome Project—One more misstep in the war on cancer. *Nature Biotechnology* 23: 535–537.

Millard, A. V., and M. A. Graham. 1985. Breastfeeding in two Mexican villages: Social and demographic perspectives. In V. Hull and M. Simpson, eds., *Breastfeeding, Child Health and Child Spacing: Cross-Cultural Perspectives*, 55–77. London: Croom Helm.

Miller, J. E., G. Rodriguez, and A. R. Pebley. 1994. Lactation, seasonality, and mother's postpartum weight change in Bangladesh: An analysis of maternal depletion. *American Journal of Human Biology* 6: 511–524.

Milman, N. 2006. Iron prophylaxis in pregnancy—General or individual and in which dose? *Annals of Hematology* 85: 821–828.

Milton, K. 2000. Hunter-gatherer diets—A different perspective. *American Journal of Clinical Nutrition* 71: 665–667.

Mitrunen, K., N. Jourenkova, V. Kataja, M. Eskelinen, V. M. Kosma, S. Benhamou, et al. 2000. Steroid metabolism gene CYP17 polymorphism and the development of breast cancer. *Cancer Epidemiology Biomarkers and Prevention* 9: 1343–1348.

Miyoshi, Y., K. Iwao, N. Ikeda, C. Egawa, and S. Noguchi. 2000. Genetic polymorphism in CYP17 and breast cancer risk in Japanese women. *European Journal of Cancer* 36: 2375–2379.

Mo, Q. H., H. Zhu, L. Y. Li, and X. M. Xu. 2004. Reliable and high-throughput mutation screening for beta-thalassemia by a single-base extension/fluorescence polarization assay. *Genetic Testing* 8: 257–262.

Møller, A. P., and J. Manning. 2003. Growth and developmental instability. *Veterinary Journal* 166: 19–27.

Moore, S. E., I. Halsall, D. Howarth, E. M. E. Poskitt, and A. M. Prentice. 2001. Glucose, insulin and lipid metabolism in rural Gambians exposed to early malnutrition. *Diabetic Medicine* 18: 646–653.

Moran, C., E. Hernandez, J. E. Ruiz, M. E. Fonseca, J. A. Bermudez, and A. Zarate. 1999. Upper body obesity and hyperinsulinemia are associated with anovulation. *Gynecologic and Obstetric Investigation* 47: 1–5.

Morgan, K. 2008. Slave women and reproduction in Jamaica, ca. 1776–1834. In G. Campbell, S. Miers, and J. C. Miller, eds., *Women and Slavery: The Modern Atlantic*, 27–53. Athens: Ohio University Press.

Morio, B., C. Montaurier, G. Pickering, P. Ritz, N. Fellmann, J. Coudert, et al. 1998. Effects of 14 weeks of progressive endurance training on energy expenditure in elderly people. *British Journal of Nutrition* 80: 511–519.

Morrison, C. D. 2008. Leptin resistance and the response to positive energy balance. *Physiology and Behavior* 94: 660–663.

Morrow, E. H., A. D. Stewart, and W. R. Rice. 2008. Assessing the extent of genome-wide intralocus sexual conflict via experimentally enforced gender-limited selection. *Journal of Evolutionary Biology* 21: 1046–1054.

287

Moussavi, M., R. Heidarpour, A. Aminorroaya, Z. Pournaghshband, and M. Amini. 2005. Prevalence of vitamin D deficiency in Isfahani high school students in 2004. *Hormone Research* 64: 144–148.

Mozaffarian, D., M. B. Katan, A. Ascherio, M. J. Stampfer, and W. C. Willett. 2006. Medical progress—Trans fatty acids and cardiovascular disease. *New England Journal of Medicine* 354: 1601–1613.

Muehlenbein, M. P., and R. G. Bribiescas. 2005. Testosterone-mediated immune functions and male life histories. *American Journal of Human Biology* 17: 527–558.

Mukamel, M. N., Y. Weisman, R. Somech, Z. Eisenberg, J. Landman, I. Shapira, et al. 2001. Vitamin D deficiency and insufficiency in Orthodox and non-Orthodox Jewish mothers in Israel. *Israel Medical Association Journal* 3: 419–421.

Muller, H. G., J. M. Chiou, J. R. Carey, and J. L. Wang. 2002. Fertility and life span: Late children enhance female longevity. *Journal of Gerontology Series A* 57: B202–206.

Murphy, M., A. Nevill, C. Neville, S. Biddle, and A. Hardman. 2002. Accumulating brisk walking for fitness, cardiovascular risk, and psychological health. *Medicine and Science in Sports and Exercise* 34: 1468–1474.

Mustillo, S., N. Krieger, E. P. Gunderson, S. Sidney, H. McCreath, and C. I. Kiefe. 2004. Self-reported experiences of racial discrimination and black-white differences in preterm and low-birthweight deliveries: The CARDIA Study. *American Journal of Public Health* 94: 2125–2131.

Myers, J., M. Prakash, V. Froelicher, D. Do, S. Partington, and J. E. Atwood. 2002. Exercise capacity and mortality among men referred for exercise testing. *New England Journal of Medicine* 346: 793–801.

Nagata, C., N. Takatsuka, S. Inaba, N. Kawakami, and H. Shimizu. 1998. Effect of soymilk consumption on serum estrogen concentrations in premenopausal Japanese women. *Journal of the National Cancer Institute* 90: 1830–1835.

National Academy of Sciences Committee on Population. 1989. *Contraception and Reproduction: Health Consequences for Women and Children in the Developing World.* Washington, D.C.: National Academy Press.

National Cancer Institute. 2012. SEER Stat Fact Sheets: Breast. Surveillance Epidemiology and End Results, http://seer.cancer.gov/statfacts/html/breast.html.

Naughton, J. M., K. O'Dea, and A. J. Sinclair. 1986. Animal foods in traditional Australian aboriginal diets: Polyunsaturated and low in fat. *Lipids* 21: 684–690.

Ness, R. B., T. Harris, J. Cobb, K. M. Flegal, J. L. Kelsey, A. Balanger, et al. 1993. Number of pregnancies and the subsequent risk of cardiovascular disease. *New England Journal of Medicine* 328: 1528–1533.

Norman, R. J., and A. M. Clark. 1998. Obesity and reproductive disorders: A review. *Reproduction Fertility and Development* 10: 55–63.

Norman, R. J., M. Noakes, R. J. Wu, M. J. Davies, L. Moran, and J. X. Wang. 2004. Improving reproductive performance in overweight/obese women with effective weight management. *Human Reproduction Update* 10: 267–280.

Ntais, C., A. Polycarpou, and J. P. A. Ioannidis. 2003. Association of the CYP17 gene polymorphism with the risk of prostate cancer: A meta-analysis. *Cancer Epidemiology Biomarkers and Prevention* 12: 120–126.

Núñez-de la Mora, A., R. T. Chatterton, O. A. Choudhury, D. A. Napolitano, and G. R. Bentley. 2007. Childhood conditions influence adult progesterone levels. *PLoS Medicine* 4: e167.

Obermeyer, C. M. 2000. Menopause across cultures: A review of the evidence. *Menopause* 7: 184–192.

O'Dea, K. 1984. Marked improvement in carbohydrate and lipid metabolism in diabetic Australian aborigines after temporary reversion to traditional life-style. *Diabetes* 33: 596–603.

———. 1991. Traditional diet and food preferences of Australian Aboriginal hunter-gatherers. *Philosophical Transactions of the Royal Society of London Series B, Biological Sciences* 334: 233–241.

Oga, T., K. Nishimura, M. Tsukino, S. Sato, and T. Hajiro. 2003. Analysis of the factors related to mortality in chronic obstructive pulmonary disease—Role of exercise capacity and health status. *American Journal of Respiratory and Critical Care Medicine* 167: 544–549.

Oguma, Y., and T. Shinoda-Tagawa. 2004. Physical activity decreases cardiovascular disease risk in women—Review and meta-analysis. *American Journal of Preventive Medicine* 26: 407–418.

Onyike, C. U., R. E. Crum, H. B. Lee, C. G. Lyketsos, and W. W. Eaton. 2003. Is obesity associated with major depression? Results from the Third National Health and Nutrition Examination Survey. *American Journal of Epidemiology* 158: 1139–1147.

Osei-Tutu, K. B., and P. D. Campagna. 2005. The effects of short- vs. long-bout exercise on mood, VO2max., and percent body fat. *Preventive Medicine* 40: 92–98.

Ounsted, M. 1986. Transmission through the female line of fetal growth constraint. *Early Human Development* 13: 339–340.

Ounsted, M., A. Scott, and C. Ounsted. 1986. Transmission through the female line of a mechanism constraining human fetal growth. *Annals of Human Biology* 13: 143–151.

Owens, J. F., K. A. Matthews, R. R. Wing, and L. H. Kuller. 1990. Physical activity and cardiovascular risk: A cross-sectional study of middle-aged premenopausal women. *Preventive Medicine* 19: 147–157.

Paffenbarger, R. S., R. T. Hyde, A. L. Wing, I. M. Lee, D. L. Jung, and J. B. Kampert. 1993. The association of changes in physical activity level and other life style characteristics with mortality among men. *New England Journal of Medicine* 328: 538–545.

Painter, R. C., S. R. de Rooij, P. M. Bossuyt, E. de Groot, W. J. Stok, C. Osmond, et al. 2007. Maternal nutrition during gestation and carotid arterial compliance in the adult offspring: The Dutch famine birth cohort. *Journal of Hypertension* 25: 533–540.

Painter, R. C., T. J. Roseboom, and O. P. Bleker. 2005. Prenatal exposure to the Dutch famine and disease in later life: An overview. *Reproductive Toxicology* 20: 345–352.

Panter-Brick, C. 1992. Working mothers in rural Nepal. In V. Maher, ed., *The Anthropology of Breast-Feeding: Natural Law or Social Construct*, 133–150. Oxford: St. Martin's Press.

———. 1993. Seasonality and levels of energy expenditure during pregnancy and lactation for rural Nepali women. *American Journal of Clinical Nutrition* 57: 620–628.

———. 1996. Physical activity, energy stores, and seasonal energy balance among men and women in Nepali households. *American Journal of Human Biology* 8: 263–274.

———. 2002. Sexual division of labor: Energetic and evolutionary scenarios. *American Journal of Human Biology* 14: 627–640.

Panter-Brick, C., D. S. Lotstein, and P. T. Ellison. 1993. Seasonality of reproductive function and weight loss in rural Nepali women. *Human Reproduction* 8: 684–690.

Papathanasiou, A., C. S. Larsen, and L. Norr. 2000. Bioarchaeological inferences from a Neolithic ossuary from Alepotrypa Cave, Diros, Greece. *International Journal of Osteoarchaeology* 10: 210–228.

Paradies, Y. 2006. A systematic review of empirical research on self-reported racism and health. *International Journal of Epidemiology* 35: 888–901.

Pawlowski, B., R. I. M. Dunbar, and A. Lipowicz. 2000. Evolutionary fitness—Tall men have more reproductive success. *Nature* 403: 156.

Peabody, J. W., P. J. Gertler, and A. Leibowitz. 1998. The policy implications of better structure and process on birth outcomes in Jamaica. *Health Policy* 43: 1–13.

Peacock, N. 1991. An evolutionary perspective on the patterning of maternal investment in pregnancy. *Human Nature* 2: 351–385.

Pedersen, S. 1993. *Family, Dependence, and the Origins of the Welfare State. Britain and France, 1914–1945.* Cambridge: Cambridge University Press.

Peterson, C. C., K. A. Nagy, and J. Diamond. 1990. Sustained metabolic scope. *Proceedings of the National Academy of Sciences of the United States of America* 87: 2324–2328.

Petit, M. A., J. C. Prior, and S. I. Barr. 1999. Running and ovulation positively change cancellous bone in premenopausal women. *Medicine and Science in Sports and Exercise* 31: 780–787.

Piers, L. S., S. N. Diggavi, S. Thangam, J. M. A. van Raaij, P. S. Shetty, and J. G. A. J. Hautvast. 1995. Changes in energy expenditure, anthropometry, and energy intake during the course of pregnancy and lactation in well-nourished Indian women. *American Journal of Clinical Nutrition* 61: 501–513.

Pike, I. L. 1999. Age, reproductive history, seasonality, and maternal body composition during pregnancy for nomadic Turkana of Kenya. *American Journal of Human Biology* 11: 658–672.

———. 2000. Pregnancy outcome for nomadic Turkana pastoralists of Kenya. *American Journal of Physical Anthropology* 113: 31–45.

Pirke, K. M., U. Schweiger, W. Lemmel, J. C. Krieg, and M. Berger. 1985. The influence of dieting on the menstrual cycle of healthy young women. *Journal of Clinical Endocrinology and Metabolism* 60: 1174–1179.

Pirke, K. M., W. Wuake, and U. Schweiger, eds. 1989. *The Menstrual Cycle and Its Disorders*. Berlin: Springer-Verlag.

Poehlman, E. T., and E. S. Horton. 1989. The impact of food intake and exercise on energy expenditure. *Nutritional Review* 47: 129–137.

Popenoe, R. 2004. *Feeding Desire: Fatness, Beauty, and Sexuality among a Saharan People*. London: Routledge.

Poppitt, S. D., A. M. Prentice, G. R. Goldberg, and R. G. Whitehead. 1994. Energy-sparing strategies to protect human fetal growth. *American Journal of Obstetrics and Gynecology* 171: 118–125.

Poppitt, S. D., A. M. Prentice, E. Jequier, Y. Schutz, and R. G. Whitehead. 1993. Evidence of energy sparing in Gambian women during pregnancy: A longitudinal study using whole-body calorimetry. *American Journal of Clinical Nutrition* 57: 353–364.

Poston, D. L., and K. B. Kramer. 1983. Voluntary and involuntary childlessness in the United States, 1955–1973. *Social Biology* 30: 290–306.

Poston, D. L., K. B. Kramer, K. Trent, and M. Y. Yu. 1983. Estimating voluntary and involuntary childlessness in the developing countries. *Journal of Biosocial Science* 15: 441–452.

Powe, C. E., C. D. Knott, and N. Conklin-Brittain. 2010. Infant sex predicts breast milk energy content. *American Journal of Human Biology* 22: 50–54.

Powys, A. O. 1905. Data for the problem of evolution in man: On fertility, duration of life, and reproductive selection. *Biometrika*: 233–285.

Prentice, A. M. 1984. Adaptations to long-term low energy intake. In E. Pollitt and P. Amante, eds., *Energy Intake and Activity*, 3–21. New York: Alan R. Liss.

———. 2000. Calcium in pregnancy and lactation. *Annual Review of Nutrition* 20: 249–272.

Prentice, A. M., T. J. Cole, F. A. Foord, W. H. Lamb, and R. G. Whitehead. 1987. Increased birthweight after prenatal dietary supplementation of rural African women. *American Journal of Clinical Nutrition* 46: 912–925.

Prentice, A. M., G. R. Goldberg, H. L. Davies, P. R. Murgatroyd, and W. Scott. 1989. Energy-sparing adaptations in human pregnancy assessed by whole-body calorimetry. *British Journal of Nutrition* 62: 5–22.

Prentice, A. M., L. M. A. Jarjou, D. M. Stirling, R. Buffenstein, and S. Fairweather-Tait. 1998. Biochemical markers of calcium and bone metabolism during 18 months of lactation in Gambian women accustomed to a low calcium intake and in those consuming a calcium supplement. *Journal of Clinical Endocrinology and Metabolism* 83: 1059–1066.

Prentice, A. M., S. D. Poppitt, G. R. Goldberg, and A. Prentice. 1995. Adaptive strategies regulating energy balance in human pregnancy. *Human Reproduction Update* 1: 149–161.

Prentice, A. M., and A. Prentice. 1990. Maternal energy requirements to support lactation. In S. A. Atkinson, L. A. Hanson, and R. K. Chandra, eds., *Breast-feeding, Nutrition, Infection and Infant Growth in Developed and Emerging Countries*, 69–86. St. John's, Newfoundland: ARTS Biomedical.

Prentice, A. M., C. J. K. Spaaij, G. R. Goldberg, S. D. Poppitt, J. M. A. van Raaij, M. Totton, et al. 1996. Energy requirements of pregnant and lactating women. *European Journal of Clinical Nutrition* 50 (Suppl.): S82–S111.

Prentice, A. M., and R. G. Whitehead. 1987. The energetics of human reproduction. *Symposia of the Zoological Society of London* 57: 275–304.

Prentice, A. M., R. G. Whitehead, S. B. Roberts, and A. A. Paul. 1981. Long-term energy balance in child-bearing Gambian women. *American Journal of Clinical Nutrition* 34: 2790–2799.

Prentice, A. M., R. G. Whitehead, M. Watkinson, W. H. Lamb, and T. J. Cole. 1983. Prenatal dietary supplementation of African women and birth-weight. *Lancet* 1: 489–492.

Prior, J. C. 1985. Luteal phase defects and anovulation: Adaptive alterations occurring with conditioning exercise. *Seminars in Reproductive Endocrinology* 3: 27–33.

———. 2007. FSH and bone—Important physiology or not? *Trends in Molecular Medicine* 13: 1–3.

Prior, J. C., K. Cameron, B. H. Yuen, and J. Thomas. 1982. Menstrual cycle changes with marathon training: Anovulation and short luteal phase. *Canadian Journal of Applied Sport Sciences* 7: 173–177.

Prior, J. C., Y. M. Vigna, and D. W. McKay. 1992. Reproduction for the athletic woman: New understandings of physiology and management. *Sports Medicine* 14: 190–199.

Qureshi, A. I., W. H. Giles, J. B. Croft, and B. J. Stern. 1997. Number of pregnancies and risk for stroke and stroke subtypes. *Archives of Neurology* 54: 203–206.

Raisz, L. G. 2005. Pathogenesis of osteoporosis: Concepts, conflicts, and prospects. *Journal of Clinical Investigation* 115: 3318–3325.

Ramakrishnan, U., H. Barnhart, D. G. Schroeder, A. D. Stein, and R. Martorell. 1999. Early childhood nutrition, education and fertility milestones in Guatemala. *Journal of Nutrition* 129: 2196–2202.

Ramakrishnan, U., R. Martorell, D. G. Schroeder, and R. Flores. 1999. Role of intergenerational effects on linear growth. *Journal of Nutrition* 129 (Suppl.): 544S–549S.

Rampersaud, E., B. D. Mitchell, T. I. Pollin, M. Fu, H. Q. Shen, J. R. O'Connell, et al. 2008. Physical activity and the association of common FTO gene variants with body mass index and obesity. *Archives of Internal Medicine* 168: 1791–1797.

Randolph, J. F., M. Sowers, E. B. Gold, B. A. Mohr, J. Luborsky, N. Santoro, et al. 2003. Reproductive hormones in the early menopausal transition: Relationship to ethnicity, body size, and menopausal status. *Journal of Clinical Endocrinology and Metabolism* 88: 1516–1522.

Rao, S., C. S. Yajnik, A. Kanade, C. H. D. Fall, B. M. Margetts, A. A. Jackson, et al. 2001. Intake of micronutrient-rich foods in rural Indian mothers is associated with the size of their babies at birth: Pune maternal nutrition study. *Journal of Nutrition* 131: 1217–1224.

Raphael, D., and F. Davis. 1985. *Only Mothers Know: Patterns of Infant Feeding in Traditional Cultures*. Westport, Conn.: Greenwood Press.

Rashid, M., and S. J. Ulijaszek. 1999. Daily energy expenditure across the course of lactation among urban Bangladeshi women. *American Journal of Physical Anthropology* 110: 457–465.

Rattan, S. I. S. 2006. Theories of biological aging: Genes, proteins, and free radicals. *Free Radical Research* 40: 1230–1238.

Redman, L. M., and A. B. Loucks. 2005. Menstrual disorders in athletes. *Sports Medicine* 35: 747–755.

Reed, T. E. 1969. Caucasian genes in American Negroes. *Science* 165: 762–768.

Reynolds, V., and R. E. S. Tanner. 1983. *The Biology of Religion*. London: Longman.

Riad-Fahmy, D., G. F. Read, and R. F. Walker. 1983. Salivary steroid assays for assessing variation in endocrine activity. *Journal of Steroid Biochemistry and Molecular Biology* 19: 265–272.

Riad-Fahmy, D., G. F. Read, R. F. Walker, S. M. Walker, and K. Griffiths. 1987. Determination of ovarian steroid hormone levels in saliva—An overview. *Journal of Reproductive Medicine* 32: 254–272.

Rice, W. R. 1998. Male fitness increases when females are eliminated from gene pool: Implications for the Y chromosome. *Proceedings of the National Academy of Sciences of the United States of America* 95: 6217–6221.

Rich-Edwards, J. W., M. B. Goldman, W. C. Willett, D. J. Hunter, M. J. Stampfer, G. A. Colditz, and J. E. Manson. 1994. Adolescent body-mass index and infertility caused by ovulatory disorder. *American Journal of Obstetrics and Gynecology* 171: 171–177.

Rich-Edwards, J. W., N. Krieger, J. Majzoub, S. Zierler, E. Lieberman, and M. Gillman. 2001. Maternal experiences of racism and violence as predictors of preterm birth: Rationale and study design. *Paediatric and Perinatal Epidemiology* 15: 124–135.

Riggs, P. 1994. The standard of living in Scotland, 1800–1850. In J. Komlos, ed., *Stature, Living Standards, and Economic Development*, 60–75. Chicago: University of Chicago Press.

Roberts, R. E., S. Deleger, W. J. Strawbridge, and G. A. Kaplan. 2003. Prospective association between obesity and depression: Evidence from the Alameda County Study. *International Journal of Obesity* 27: 514–521.

Roberts, R. E., G. A. Kaplan, S. J. Shema, and W. J. Strawbridge. 2000. Are obese at greater risk for depression? *American Journal of Epidemiology* 152: 163–170.

Roberts, S. B., A. A. Paul, T. J. Cole, and R. G. Whitehead. 1982. Seasonal changes in activity, birth weight and lactational performance in rural Gambian women. *Transactions of the Royal Society of Tropical Medicine and Hygiene* 76: 668–678.

Robertson, C. 1996. Africa into the Americas? Slavery and women, the family, and the gender division of labor. In D. B. Gaspar and D. Clark Hine, eds., *More Than Chattel: Black Women and Slavery in the Americas*, 3–40. Bloomington: Indiana University Press.

Rock, C. L., S. W. Flatt, C. A. Thomson, M. L. Stefanick, V. A. Newman, L. A. Jones, et al. 2004. Effects of a high-fiber, low-fat diet intervention on serum concentrations of reproductive steroid hormones in women with a history of breast cancer. *Journal of Clinical Oncology* 22: 2379–2387.

Rockhill, B., W. C. Willett, D. J. Hunter, J. E. Manson, S. E. Hankinson, and G. A. Colditz. 1999. A prospective study of recreational physical activity and breast cancer risk. *Archives of Internal Medicine* 159: 2290–2296.

Rogers, M. E., F. Maisels, E. A. Williamson, M. Fernandez, and C. E. G. Tutin. 1990. Gorilla diet in the Lope Reserve, Gabon—A nutritional analysis. *Oecologia* 84: 326–339.

Rosamond, W., K. Flegal, G. Friday, K. Furie, A. Go, K. Greenlund, et al. 2007. Heart disease and stroke statistics—2007 update: A report from the American Heart Association Statistics Committee and Stroke Statistics Subcommittee. *Circulation* 115: E69–E171.

Rose, D. P., M. Lubin, and J. M. Connolly. 1997. Effects of diet supplementation with wheat bran on serum estrogen levels in the follicular and luteal phases of the menstrual cycle. *Nutrition* 13: 535–539.

Rose, J. C. 1989. Biological consequences of segregation and economic deprivation—A post-slavery population from southwest Arkansas. *Journal of Economic History* 49: 351–360.

Roseboom, T., S. de Rooij, and R. Painter. 2006. The Dutch famine and its long-term consequences for adult health. *Early Human Development* 82: 485–491.

Rosenberg, L., J. R. Palmer, L. A. Wise, N. J. Horton, and M. J. Corwin. 2002. Perceptions of racial discrimination and the risk of preterm birth. *Epidemiology* 13: 646–652.

Rosetta, L. 1993. Female reproductive dysfunction and intense physical training. *Oxford Reviews in Reproductive Biology* 15: 113–141.

———. 2002. Female fertility and intensive physical activity. *Science and Sports* 17: 269–277.

Rosetta, L., G. A. Harrison, and G. F. Read. 1998. Ovarian impairments of female recreational distance runners during a season of training. *Annals of Human Biology* 25: 345–357.

Ross, R. K., A. Paganini-Hill, P. C. Wan, and M. C. Pike. 2000. Effect of hormone replacement therapy on breast cancer risk: Estrogen versus estrogen plus progestin. *Journal of the National Cancer Institute* 92: 328–332.

Rossouw, J. E., G. L. Anderson, R. L. Prentice, A. Z. LaCroix, C. Kooperberg, M. L. Stefanick, et al. 2002. Risks and benefits of estrogen plus progestin in healthy postmenopausal women—Principal results from the Women's Health Initiative randomized controlled trial. *Journal of the American Medical Association* 288: 321–333.

Roumen, F. J. M. E., W. H. Doesburg, and R. Rolland. 1982. Hormonal patterns in infertile women with a deficient postcoital test. *Fertility and Sterility* 38: 24–47.

Russell, R. J. H., and P. A. Wells. 1987. Estimating paternity confidence. *Ethology and Sociobiology* 8: 215–220.

Saadi, H. F., N. Nagelkerke, B. Sheela, H. S. Qazaq, E. Zilahi, M. K. Mohamadiyeh, and A. I. Al-Suhaili. 2006. Predictors and relationships of serum 25 hydroxyvitamin D concentration with bone turnover markers, bone mineral density, and vitamin D receptor genotype in Emirati women. *Bone* 39: 1136–1143.

Sachan, A., R. Gupta, V. Das, A. Agarwal, P. K. Awasthi, and V. Bhatia. 2005. High prevalence of vitamin D deficiency among pregnant women and their newborns in northern India. *American Journal of Clinical Nutrition* 81: 1060–1064.

Salmon, A. B., D. B. Marx, and L. G. Harshman. 2001. A cost of reproduction in *Drosophila melanogaster:* Stress susceptibility. *Evolution* 55: 1600–1608.

Santoro, N., L. T. Goldsmith, D. Heller, N. Illsley, P. McGovern, C. Molina, et al. 2000. Luteal progesterone relates to histological endometrial maturation in fertile women. *Journal of Clinical Endocrinology and Metabolism* 85: 4207–4211.

Saris, W. H. M., M. A. Vanerpbaart, F. Brouns, K. R. Westerterp, and F. ten Hoor. 1989. Study on food intake and energy expenditure during extreme sustained exercise—The Tour de France. *International Journal of Sports Medicine* 10: S26–S31.

Savitt, T. L. 1978. *Medicine and Slavery. The Diseases and Health Care of Blacks in Antebellum Virginia.* Urban: University of Illinois Press.

Schmitt, E., and H. Stopper. 2001. Estrogenic activity of naturally occurring anthocyanidins. *Nutrition and Cancer* 41: 145–149.

Schweiger, U., R. J. Tuschl, R. G. Laessle, A. Broocks, and K. M. Pirke. 1989. Consequences of dieting and exercise on menstrual function in normal weight women. In K. M. Pirke, W. Wuttke, and U. Schweiger, eds., *The Menstrual Cycle and Its Disorders,* 142–149. Berlin: Springer-Verlag.

Sear, R., R. Mace, and I. A. McGregor. 2000. Maternal grandmothers improve nutritional status and survival of children in rural Gambia. *Proceedings of the Royal Society of London B, Biological Sciences* 267: 1641–1647.

Seghieri, G., A. De Bellis, R. Anichini, L. Alviggi, F. Franconi, and M. C. Breschi. 2005. Does parity increase insulin resistance during pregnancy? *Diabetic Medicine* 22: 1574–1580.

Sellen, D. W. 2000. Seasonal ecology and nutritional status of women and children in a Tanzanian pastoral community. *American Journal of Human Biology* 12: 758–781.

Selling, K. E., J. Carstensen, O. Finnstrom, and G. Sydsjo. 2006. Intergenerational effects of preterm birth and reduced intrauterine growth: A population-based study of Swedish mother-offspring pairs. *BJOG* 113: 430–440.

Setchell, K. D. R. 2001. Soy isoflavones—Benefits and risks from nature's selective estrogen receptor modulators (SERMs). *Journal of the American College of Nutrition* 20 (Suppl.): 354S–362S.

Setchell, K. D. R., and A. Cassidy. 1999. Dietary isoflavones: Biological effects and relevance to human health. *Journal of Nutrition* 129: 758S–767S.

Setchell, K. D. R., and E. Lydeking-Olsen. 2003. Dietary phytoestrogens and their effect on bone: Evidence from in vitro and in vivo, human observational, and dietary intervention studies. *American Journal of Clinical Nutrition* 78: 593S–609S.

Sharp, J. 1671. *The Midwives Book, or the Whole Art of Midwifery Discovered, Directing Childbearing Women How to Behave Themselves in Their Conception, Breeding, Bearing, and Nursing of Children*. London: Simon Miller.

Shea, J. L. 2006. Cross-cultural comparison of women's midlife symptom-reporting: A China study. *Culture Medicine and Psychiatry* 30: 331–362.

Shell-Duncan, B., and S. A. Yung. 2004. The maternal depletion transition in northern Kenya: The effects of settlement, development and disparity. *Social Science and Medicine* 58: 2485–2498.

Shephard, R. J. 1980. Work physiology and activity patterns. In F. A. Milan, ed., *The Human Biology of Circumpolar Populations*, 305–338. Cambridge: Cambridge University Press.

Sherman, P. W. 1998. Animal behavior—The evolution of menopause. *Nature* 392: 759–761.

Shirtcliff, E. A., D. A. Granger, E. B. Schwartz, M. J. Curran, A. Booth, and W. H. Overman. 2000. Assessing estradiol in biobehavioral studies using saliva and blood spots: Simple radioimmunoassay protocols, reliability, and comparative validity. *Hormones and Behavior* 38: 137–147.

Shostak, M. 1983. *Nisa: The Life and Words of a !Kung Woman*. New York: Vintage Books.

Simmons, D. 1992. Parity, ethnic group and the prevalence of type-2 diabetes—The Coventry Diabetes Study. *Diabetic Medicine* 9: 706–709.

Simmons, D., J. Shaw, A. McKenzie, S. Eaton, A. J. Cameron, and P. Zimmet. 2006. Is grand multiparity associated with an increased risk of dysglycaemia? *Diabetologia* 49: 1522–1527.

Simpson, E. R. 2002. Aromatization of androgens in women: Current concepts and findings. *Fertility and Sterility* 77: S6–S10.

Simpson, E. R., C. Clyne, G. Rubin, W. C. Boon, K. Robertson, K. Britt, et al. 2002. Aromatase—A brief overview. *Annual Review of Physiology* 64: 93–127.

Sinclair, A. J., and K. O'Dea. 1990. Fats in human diets through history: Is the Western diet out of step? In J. D. Wood and A. V. Fisher, eds., *Reducing Fat in Meat Animals*, 1–47. London: Elsevier Applied Science.

Singh, J., A. M. Prentice, E. Diaz, W. A. Coward, J. Ashford, M. Sawyer, and R. G. Whitehead. 1989. Energy expenditure of Gambian women during peak agricultural activity measured by the doubly-labelled water method. *British Journal of Nutrition* 62: 315–329.

Sjodin, A. M., A. H. Forslund, K. R. Westerterp, A. B. Andersson, J. W. Forslund, and L. H. Hammbraeus. 1996. The influence of physical activity on BMR. *Medicine and Science in Sports and Exercise* 28: 85–91.

Small, C. M., M. Marcus, S. L. Sherman, A. K. Sullivan, A. K. Manatunga, and H. S. Feigelson. 2005. CYP17 genotype predicts serum hormone levels among pre-menopausal women. *Human Reproduction* 20: 2162–2167.

Smith, A. H., T. M. Butler, and N. Pace. 1975. Weight growth of colony-reared chimpanzees. *Folia Primatologica* 24: 29–59.

Sobo, E. 1993. *One Blood: The Jamaican Body.* Albany: State University of New York Press.

Somner, J., S. McLellan, J. Cheung, Y. T. Mak, M. L. Frost, K. M. Knapp, et al. 2004. Polymorphisms in the p450 c17 (17-hydroxylase/17,20-lyase) and p450 c19 (aromatase) genes: Association with serum sex steroid concentrations and bone mineral density in postmenopausal women. *Journal of Clinical Endocrinology and Metabolism* 89: 344–351.

Sorenson Jamison, C., P. L. Jamison, and L. L. Cornell. 2005. Human female longevity: How important is being a grandmother? In E. Voland, A. Chasiotis, and W. Schiefenhovel, eds., *Grandmotherhood: The Evolutionary Significance of the Second Half of Female Life,* 99–117. New Brunswick, N.J.: Rutgers University Press.

Soto, A. M., and C. Sormenschein. 1987. Cell proliferation of estrogen sensitive cells: The case for negative control. *Endocrine Reviews* 8: 44–52.

Spencer, C. P., E. P. Morris, and J. M. Rymer. 1999. Selective estrogen receptor modulators: Women's panacea for the next millennium? *American Journal of Obstetrics and Gynecology* 180: 763–770.

Spurdle, A. B., J. L. Hopper, G. S. Dite, X. Chen, J. Cui, M. R. McCredie, et al. 2000. CYP17 promoter polymorphism and breast cancer in Australian women under age forty years. *Journal of the National Cancer Institute* 92: 1674–1681.

Stanford, C. B. 1998. *Chimpanzee and Red Colobus: The Ecology of Predator and Prey.* Cambridge, Mass.: Harvard University Press.

Stanford, J. L., E. A. Noonan, L. Iwasaki, S. Kolb, R. B. Chadwick, Z. D. Feng, and E. A. Ostrander. 2002. A polymorphism in the CYP17 gene and risk of prostate cancer. *Cancer Epidemiology Biomarkers and Prevention* 11: 243–247.

Stearns, S. C., R. M. Nesse, and D. Haig. 2008. Introducing evolutionary thinking for medicine. In S. C. Stearns and J. C. Koella, eds., *Evolution in Health and Disease,* 3–14. New York: Oxford University Press.

Steckel, R. H. 1986a. Birth weights and infant mortality among American slaves. *Explorations in Economic History* 23: 173–198.

———. 1986b. A dreadful childhood—The excess mortality of American slaves. *Social Science History* 10: 427–465.

———. 1994. Heights and health in the United States, 1710–1950. In J. Komlos, ed., *The Height of Runaway Slaves in Colonial America, 1720–1770,* 153–170. Chicago: University of Chicago Press.

———. 1996. Women, work, and health under plantation slavery in the United States. In D. B. Gaspar and D. Clark Hine, eds., *More Than Chattel: Black Women and Slavery in the Americas,* 43–60. Bloomington: Indiana University Press.

Stein, A. D., and L. H. Lumey. 2000. The relationship between maternal and offspring birth weights after maternal prenatal famine exposure: The Dutch Famine Birth Cohort Study. *Human Biology* 72: 641–654.

Sterkowicz, S. 1952a. Krakowska Kropla Mleka, part 1. *Polski Tygodnik Lekarski* 14: 413–415.

———. 1952b. Krakowska Kropla Mleka, part 2. *Polski Tygodnik Lekarski* 15: 450–454.

Sternfeld, B. 1997. Physical activity and pregnancy outcome—Review and recommendations. *Sports Medicine* 23: 33–47.

Stoddard, J. L., C. W. Dent, L. Shames, and L. Bernstein. 2007. Exercise training effects on premenstrual distress and ovarian steroid hormones. *European Journal of Applied Physiology* 99: 27–37.

Strassmann, B. I. 1996. Energy economy in the evolution of menstruation. *Evolutionary Anthropology* 5: 157–164.

———. 1997a. The biology of menstruation in *Homo sapiens:* Total lifetime menses, fecundity, and nonsynchrony in a natural-fertility population. *Current Anthropology* 38: 123–129.

———. 1997b. Polygyny as a risk factor for child mortality among the Dogon. *Current Anthropology* 38: 688–695.

Strassmann, B. I., and B. Gillespie. 2002. Life-history theory, fertility and reproductive success in humans. *Proceedings of the Royal Society of London Series B, Biological Sciences* 269: 553–562.

Strassmann, B. I., and R. Mace. 2008. Perspectives on human health and disease from evolutionary and behavioral ecology. In S. C. Stearns and J. C. Koella, eds., *Evolution in Health and Disease,* 109–121. New York: Oxford University Press.

Streeten, E. A., K. A. Ryan, D. J. McBride, T. I. Pollin, A. R. Shuldiner, and B. D. Mitchell. 2005. The relationship between parity and bone mineral density in women characterized by a homogeneous lifestyle and high parity. *Journal of Clinical Endocrinology and Metabolism* 90: 4536–4541.

Suarez, R. K. 1996. Upper limits to mass-specific metabolic rates. *Annual Review of Physiology* 58: 583–605.

Sukalich, S., S. F. Lipson, and P. T. Ellison. 1994. Intra and interwomen variation in progesterone profiles. *American Journal of Physical Anthropology* 18 (Suppl.): 191.

Swallow, D. M. 2003. Genetics of lactase persistence and lactose intolerance. *Annual Review of Genetics* 37: 197–219.

Swan, D. E. 1972. *The Structure and Profitability of the Antebellum Industry, 1859.* New York: Arno Press.

Taaffle, D. R., L. Pruitt, J. Reim, G. Butterfield, and R. Marcus. 1995. Effect of sustained resistance training on basal metabolic rate in older women. *Journal of American Geriatric Society* 43: 465–471.

Tafari, N., R. L. Naeye, and A. Gobezie. 1980. Effects of maternal undernutrition and heavy physical work during pregnancy on birth weight. *British Journal of Obstetrics and Gynaecology* 87: 222–226.

Tanasescu, N., M. F. Leitzmann, E. B. Rimm, W. C. Willett, M. J. Stampfer, and F. B. Hu. 2002. Exercise type and intensity in relation to coronary heart disease in men. *Journal of the American Medical Association* 288: 1994–2000.

Techatraisak, K., G. S. Conway, and G. Rumsby. 1997. Frequency of a polymorphism in the regulatory region of the 17 alpha-hydroxylase-17,20-lyase (CYP17) gene in hyperandrogenic states. *Clinical Endocrinology* 46: 131–134.

Thame, M., R. J. Wilks, N. McFarlane-Anderson, F. I. Bennett, and T. E. Forrester. 1997. Relationship between maternal nutritional status and infant's weight and body proportions at birth. *European Journal of Clinical Nutrition* 51: 134–138.

Thompson, K., R. Morley, S. R. Grover, and M. R. Zacharin. 2004. Postnatal evaluation of vitamin D and bone health in women who were vitamin D-deficient in pregnancy, and in their infants. *Medical Journal of Australia* 181: 486–488.

Thong, Y. H., R. W. Steele, M. M. Vincent, S. A. Hensen, and J. A. Bellanti. 1973. Impaired in vitro cell-mediated immunity to Rubella-virus during pregnancy. *New England Journal of Medicine* 289: 604–606.

Thune, I. 2000. Assessments of physical activity and cancer risk. *European Journal of Cancer Prevention* 9: 387–393.

Thune, I., T. Brenn, E. Lund, and M. Gaard. 1997. Physical activity and the risk of breast cancer. *New England Journal of Medicine* 336: 1269–1275.

Thune, I., and A. S. Furberg. 2001. Physical activity and cancer risk: Dose-response and cancer, all sites and site-specific. *Medicine and Science in Sports and Exercise* 33: S530–S550.

Tietjen, A. M. 1985. Infant care and feeding practices and beginning of socialization among the Maisin of Papua New Guinea. In L. B. Marshall, ed., *Infant Care and Feeding in the South Pacific,* 121–136. New York: Gordon and Breach Science Publishers.

Tillyard, S. 1994. *Aristocrats: Caroline, Emily, Louisa, and Sarah Lennox, 1740–1832.* New York: Farrar, Straus and Giroux.

Timar, O., F. Sestier, and E. Levy. 2000. Metabolic syndrome X: A review. *Canadian Journal of Cardiology* 16: 779–789.

Toescu, V., S. L. Nuttall, U. Martin, M. J. Kendall, and F. Dunne. 2002. Oxidative stress and normal pregnancy. *Clinical Endocrinology* 57: 609–613.

Tracer, D. P. 1991. Fertility-related changes in maternal body composition among the Au of Papua New Guinea. *American Journal of Physical Anthropology* 85: 393–406.

Travis, R. C., M. Churchman, S. A. Edwards, G. Smith, P. K. Verkasalo, C. R. Wolf, et al. 2004. No association of polymorphisms in CYP17, CYP19, and HSD17-B1 with plasma estradiol concentrations in 1,090 British women. *Cancer Epidemiology Biomarkers and Prevention* 13: 2282–2284.

Trayhurn, P. 1989. Thermogenesis and the energetics of pregnancy and lactation. *Canadian Journal of Physiology and Pharmacology* 67: 370–375.

Tremblay, A., E. T. Poehlman, J. P. Despres, G. Theriault, E. Danforth, and C. Bouchard. 1997. Endurance training with constant energy intake in identical twins: Changes over time in energy expenditure and related hormones. *Metabolism: Clinical and Experimental* 46: 499–503.

Tretli, S. 1989. Height and weight in relation to breast cancer morbidity and mortality: A prospective study of 570,000 women in Norway. *International Journal of Cancer* 44: 23–30.

Tretli, S., and M. Gaard. 1996. Lifestyle changes during adolescence and risk of breast cancer: An ecology study of the effect of World War II in Norway. *Cancer Causes and Control* 7: 507–512.

Trichopoulos, D. 1990. Hypothesis: Does breast cancer originate in utero? *Lancet* 335: 939–940.

Trivers, R. L. 1974. Parent-offspring conflict. *American Zoologist* 14: 247–262.

Trussell, J., and R. H. Steckel. 1978. Age of slaves at menarche and their 1st birth. *Journal of Interdisciplinary History* 8: 477–505.

Tsuya, N. O., S. Kurosu, and H. Nakazato. 2004. Mortality and household in two Ou villages 1716–1870. In T. Bengtsson, C. Cameron, and J. Z. Lee, eds., *Life under Pressure: Mortality and Living Standards in Europe and Asia, 1700–1900*, 253–292. Cambridge, Mass.: MIT Press.

Turner, M. 2002. "The 11 o'clock flog": Women, work and labour law in the British Caribbean. In V. A. Shepherd, ed., *Working Slavery, Pricing Freedom: Perspectives from the Caribbean, Africa and the African Diaspora*, 249–272. Kingston: Ian Randle.

Ungar, P. S., and M. F. Teaford. 2002. Perspectives on the evolution of human diet. In P. S. Ungar and M. F. Teaford, eds., *Human Diet: Its Origin and Evolution*, 1–6. Westport, Conn.: Bergin & Garvey.

Valeggia, C. R., and P. T. Ellison. 2003. Impact of breastfeeding on anthropometric changes in peri-urban Toba women (Argentina). *American Journal of Human Biology* 15: 717–724.

———. 2004. Lactational amenorrhoea in well-nourished Toba women of Formosa, Argentina. *Journal of Biosocial Science* 36: 573–595.

Van de Putte, B., K. Matthijs, and R. Vlietinck. 2003. A social component in the negative effect of sons on maternal longevity in pre-industrial humans. *Journal of Biosocial Science* 36: 289–297.

Van Zant, R. S. 1992. Influence of diet and exercise on energy expenditure—A review. *International Journal of Sport Nutrition* 2: 1–19.

Venners, S. A., X. Liu, M. J. Perry, S. A. Korrick, Z. P. Li, F. Yang, et al. 2006. Urinary estrogen and progesterone metabolite concentrations in menstrual cycles of fertile women with non-conception, early pregnancy loss or clinical pregnancy. *Human Reproduction* 21: 2272–2280.

Verkasalo, P. K., H. V. Thomas, P. N. Appleby, G. K. Davey, and T. J. Key. 2001. Circulating levels of sex hormones and their relation to risk factors for breast cancer: A cross-sectional study in 1092 pre- and postmenopausal women (United Kingdom). *Cancer Causes and Control* 12: 47–59.

Vermeulen, A. 1993. Environment, human reproduction, menopause, and andropause. *Environmental Health Perspectives* 2: 91–100.

Vigersky, R. A., A. E. Anderson, R. H. Thompson, and D. L. Loriaux. 1977. Hypothalamic dysfunction in secondary amenorrhea associated with simple weight loss. *New England Journal of Medicine* 297: 1141–1145.

Vihko, R., and D. Apter. 1984. Endocrine characteristics of adolescent menstrual cycles—Impact of early menarche. *Journal of Steroid Biochemistry and Molecular Biology* 20: 231–236.

Vitzthum, V. J. 2001. Why not so great is still good enough: Flexible responsiveness in human reproductive functioning. In P. T. Ellison, ed., *Reproductive Ecology and Human Evolution*, 179–202. New York: Aldine de Gruyter.

Vitzthum, V. J., and K. Ringheim. 2005. Hormonal contraception and physiology: A research-based theory of discontinuation due to side effects. *Studies in Family Planning* 36: 13–32.

Vitzthum, V. J., H. Spielvogel, and J. Thornburg. 2004. Interpopulational differences in progesterone levels during conception and implantation in humans. *Proceedings of the National Academy of Sciences of the United States of America* 101: 1443–1448.

Voland, E., and J. Beise. 2002. Opposite effects of maternal and paternal grandmothers on infant survival in historical Krummhorn. *Behavioral Ecology and Sociobiology* 52: 435–443.

Wadelius, M., S. O. Andersson, J. E. Johansson, C. Wadelius, and A. Rane. 1999. Prostate cancer associated with CYP17 genotype. *Pharmacogenetics* 9: 635–639.

Wallace, J. M., R. P. Aitken, J. S. Milne, and W. W. Hay. 2004. Nutritionally mediated placental growth restriction in the growing adolescent: Consequences for the fetus. *Biology of Reproduction* 71: 1055–1062.

Wall-Scheffler, C. M., K. Geiger, and K. L. Steudel-Numbers. 2007. Infant carrying: The role of increased locomotory costs in early tool development. *American Journal of Physical Anthropology* 133: 841–846.

Wappner, R., S. C. Cho, R. A. Kronmal, V. Schuett, and M. R. Seashore. 1999. Management of phenylketonuria for optimal outcome: A review of guidelines for phenylketonuria management and a report of surveys of parents, patients, and clinic directors. *Pediatrics* 104: e68.

Warren, M. P. 1990. Weight control. *Seminars in Reproductive Endocrinology* 8: 25–31.

Weiner, J. 1989. Metabolic constraints to mammalian energy budgets. *Acta Theriologica* 34: 3–36.

———. 1992. Physiological limits to sustainable energy budgets in birds and mammals: Ecological implications. *Trends in Ecology and Evolution* 7: 384–388.

Wells, A. V. 1963. Study of birth weights of babies born in Barbados, West Indies. *West Indian Medical Journal* 12: 194–199.

Wells, J. C. K. 2003. The thrifty phenotype hypothesis: Thrifty offspring or thrifty mother? *Journal of Theoretical Biology* 221: 143–161.

Westendorp, R. G. J., and T. B. L. Kirkwood. 1998. Human longevity at the cost of reproductive success. *Nature* 396: 743–746.

Weston, A., C. F. Pan, I. J. Bleiweiss, H. B. Ksieski, N. Roy, N. Maloney, and M. S. Wolff. 1998. CYP17 genotype and breast cancer risk. *Cancer Epidemiology Biomarkers and Prevention* 7: 941–944.

Whiting, M. G. 1958. A cross-cultural nutrition survey. PhD diss., School of Public Health, Harvard University.

Wiebe, J. P. 2006. Progesterone metabolites in breast cancer. *Endocrine-Related Cancer* 13: 717–738.

Willett, W. C., and M. J. Stampfer. 2003. Rebuilding the food pyramid. *Scientific American* 288: 64–71.

Williams, G. C., and R. M. Nesse. 1991. The dawn of Darwinian medicine. *Quarterly Review of Biology* 66: 1–22.

Winkvist, A., J. P. Habicht, and K. M. Rasmussen. 1998. Linking maternal and infant benefits of a nutritional supplement during pregnancy and lactation. *American Journal of Clinical Nutrition* 68: 656–661.

Winkvist, A., K. M. Rasmussen, and J. P. Habicht. 1992. A new definition of Maternal Depletion Syndrome. *American Journal of Public Health* 82: 691–694.

Winters, K. M., W. C. Adams, C. N. Meredith, M. D. VanLoan, and B. L. Lasley. 1996. Bone density and cyclic ovarian function in trained runners and active controls. *Medicine and Science in Sports and Exercise* 28: 776–785.

Wolf, J. H. 2001. *Don't Kill Your Baby: Public Health and the Decline of Breastfeeding in the Nineteenth and Twentieth Centuries.* Columbus: Ohio State University Press.

Woolf, A. D. 2006. The global perspective of osteoporosis. *Clinical Rheumatology* 25: 613–618.

Wrangham, R. W., and N. Conklin-Brittain. 2003. Cooking as a biological trait. *Comparative Biochemistry and Physiology—Part A: Molecular and Integrative Physiology* 136: 35–46.

Wrangham, R. W., J. H. Jones, G. Laden, D. Pilbeam, and N. Conklin-Brittain. 1999. The raw and the stolen—Cooking and the ecology of human origins. *Current Anthropology* 40: 567–594.

Wu, A. H., A. Seow, K. Arakawa, D. Van Den Berg, H. P. Lee, and M. C. Yu. 2003. HSD17B1 and CYP17 polymorphisms and breast cancer risk among Chinese women in Singapore. *International Journal of Cancer* 104: 450–457.

Wu, J. 1994. How severe was the Great Depression? Evidence from the Pittsburgh region. In J. Komlos, ed., *Stature, Living Standards, and Economic Development,* 129–152. Chicago: University of Chicago Press.

Wu, L. L. 1999. Review of risk factors for cardiovascular diseases. *Annals of Clinical and Laboratory Science* 29: 127–133.

Xu, X., A. M. Duncan, B. E. Merz, and M. S. Kurzer. 1998. Effects of soy isoflavones on estrogen and phytoestrogen metabolism in premenopausal women. *Cancer Epidemiology Biomarkers and Prevention* 7: 1101–1108.

Yetman, N. R., ed. 1999. *Voices from Slavery: 100 Authentic Slave Narratives.* Mineola, N.Y.: Dover Publications.

Young, I. E., K. M. Kurian, C. Annink, I. H. Kunkler, V. A. Anderson, B. B. Cohen, et al. 1999. A polymorphism in the CYP17 gene is associated with male breast cancer. *British Journal of Cancer* 81: 141–143.

Yu, M. W., Y. C. Yang, S. Y. Yang, S. W. Cheng, Y. F. Liaw, S. M. Lin, and C. J. Chen. 2001. Hormonal markers and hepatitis B virus-related hepatocellular carcinoma risk: A nested case-control study among men. *Journal of the National Cancer Institute* 93: 1644–1651.

Zaadstra, B. M., J. C. Seidell, P. A. H. Van Noord, E. R. Tevelde, J. D. F. Habbema, B. Vrieswijk, and J. Karbaat. 1993. Fat and female fecundity—prospective study of effect of body fat distribution on conception rates. *British Medical Journal* 306: 484–487.

Zelnik, M. 1966. Fertility of American Negro in 1830 and 1850. *Population Studies* 20: 77–83.

Zera, A. J., and L. G. Harshman. 2001. The physiology of life history trade-offs in animals. *Annual Reviews of Ecology and Systematics* 32: 95–126.

Ziomkiewicz, A. 2006. Anthropometric correlates of the concentration of progesterone and estradiol in menstrual cycles of women age 24–37 living in rural and urban area of Poland [in Polish]. PhD diss., Jagiellonian University, Krakow.

Ziomkiewicz, A., P. T. Ellison, S. F. Lipson, I. Thune, and G. Jasienska. 2008. Body fat, energy balance and estradiol levels: A study based on hormonal profiles from complete menstrual cycles. *Human Reproduction* 23: 2555–2563.

Zittermann, A. 2003. Vitamin D in preventive medicine: Are we ignoring the evidence? *British Journal of Nutrition* 89: 552–572.

Zmuda, J. M., J. A. Cauley, L. H. Kuller, and R. E. Ferrell. 2001. A common promoter variant in the cytochrome P450c17 alpha (CYP17) gene is associated with bioavailable testosterone levels and bone size in men. *Journal of Bone and Mineral Research* 16: 911–917.

ACKNOWLEDGMENTS

It is hard to determine how far back we need to go when assessing the influence of our past experiences on the way we think today. I will undoubtedly omit many people from my past who should be thanked in this book.

As far as early experiences go, I am very grateful to my father, who never censored my reading lists. Not even once as a child did I hear from him that a book was too serious for me or that I was too young for a particular book. He left it to me to decide if what I was reading was appropriate for my age, and this strategy worked well as I often dropped books I could not understand after reading a chapter or two. But I read many "non-children's" books when I was a child. Reading serious books so early in life was probably important for the intellectual development of a girl growing up in a small provincial town in communist Poland. It was always clear to both my parents that I should go to college, even though it was neither obvious nor easy for many young people or their parents who lived in small towns.

I owe my first encounter with evolutionary biology to Adam Lomnicki at Jagiellonian University in Krakow, my alma mater; every Monday at 6:00 PM he would hold seminars where students and some faculty discussed the most influential papers from evolutionary biology. During these Mondays, the late Jan Rafinski (my master's thesis advisor), Adam Lomnicki, Jan Kozlowski, Jacek Szumura, and January Weiner greatly affected my "Darwinian" development.

I would never have become a scientist, though, were it not for Peter T. Ellison from the Department of Biological Anthropology at Harvard University. Peter had, and still has, the most contagious enthusiasm for science. Just as anyone who was lucky enough to interact with Peter, I became deeply influenced by his way of thinking—his attention to detail and his ability to see not only the big picture but also new and original ways of looking at problems.

Peter was not afraid to accept as his graduate student a woman who came from behind the Iron Curtain with a small child, poor English, and a basically unverifiable educational background. His lab, weekly reading groups, and informal conversations provided the most stimulating of intellectual atmospheres. I also learned a lot from interactions with people who over the years were members of Peter's lab—Susan Lipson, Gillian Bentley, Rick Bribiescas, Pippi Ellison, Judith Flynn, Cheryl Knott, Cathy Lager, Mary O'Rourke, and Diana Sherry, as well as other graduate students in the department—Babette Fahey, John Kingston, Dan Lieberman, Laura MacLatchy, and many others.

Faculty at the Anthropology Department enriched my way of thinking in areas other than human reproductive ecology. I owe a lot to Irvin DeVore, David Pilbeam, Maryellen Ruvolo, Richard Wrangham, and Ofer Bar-Yosef.

My students and colleagues in Poland also have helped with my research (the results of which are described in this book), and they continue to help with studies we are currently conducting. Without them, none of these studies would have been possible. I am especially grateful to Andrzej Galbarczyk, Magdalena Klimek, Anna Merklinger-Gruchala, Ilona Nenko, and Anna Ziomkiewicz-Wichary.

I would have never begun to write this book if it were not for the 2005–2006 fellowship from the Radcliffe Institute of Advanced Study. Thanks to their support, I had the chance to spend a year in a nourishing environment, where I had time to think and research areas unrelated to my specialty. I am very grateful to Judith Vichniac, whom I admired for creating such a warm, comfortable, and incredibly stimulating environment. I would also like to thank Drew G. Faust, Barbara J. Grosz, and Louise Richardson and the fellowship staff, especially Melissa Synnott, Tony Rufo, and Sophia Heller. During that year I also interacted with other people at Radcliffe, and I was especially impressed by the professionalism of Tamara Rogers and Diane Mercer. It was a great pleasure to interact with everyone who worked at Radcliffe during that year.

I was incredibly privileged to get to know the other fellows at Radcliffe. My work has benefited greatly from historians, the specialists in the area of slavery who guided me through the unfamiliar area of research. A talk that Vincent Brown gave on mortality in Jamaican slave populations inspired me to think about a comparable approach to these aspects of slavery from my own perspective. Vincent suggested many sources, especially on Caribbean slavery. Tera Hunter directed me to data on slave children, and Eve Troutt Powell encouraged me to pursue my research.

Tony Horwitz—author, journalist, and specialist on Captain Cook and the American Civil War—also has a brilliant understanding of human biology and inspired me to think more deeply about many problems, ranging from the biological conditions of slavery to the maternal depletion syndrome. Someday, maybe I will write that trashy book we talked about (I don't mean this one). Linguist Geoffrey K. Pullum gave me the confidence to ignore some suggestions from my spelling and grammar software, which always corrects my "which." Kind interest from the late philosopher Barbara C. Scholz in my very undeveloped ideas convinced me that pursuing the relationship between slavery and low birth weight in African Americans might be worth my efforts.

Two of the talks that I gave during that year, Radcliffe Seminar and Radcliffe Breakfast in New York, both mostly to nonspecialists, allowed me to interact with the most interested and intellectually stimulating audiences I have ever met.

Peter T. Ellison, Gillian Bentley, and Richard Bribiescas were the first reviewers of my book proposal. I am grateful that they found the initial book idea interesting, and I hope that they will not be disappointed that the final product goes in some new and different directions from those that were planned. Gillian also had many detailed, very useful comments on the book manuscript—I value her advice greatly. The anonymous reviewers suggested important finishing touches. Cassandra Niemi worked as my research assistant, and it was a pleasure to interact with her. My mother-in-law Katarzyna Zachwatowicz-Jasienska inspired me to look at early twentieth-century child-welfare programs in Poland and helped find the materials. When I had doubts as to whether I should continue writing, Maria Kapiszewska gave me the most convincing reason why I should keep on.

My son Adam has always thought that his mother does the coolest things, and I constantly get my batteries recharged by his love and support.

And how to thank the person without whom this book would not be possible at all? I would quickly go beyond the word limit specified by my book contract if I were to try to list just some of those things, so I will just say—thank you, Michał.

Note: Page numbers followed by *f* and *t* indicate figures and tables.